The Subterranean Forest

THE SUBTERRANEAN FOREST

Energy Systems and the Industrial Revolution

Rolf Peter Sieferle

The White  Horse Press

333.79094
S57a

Published in German in 1982 as *Der unterirdische Wald* by
C.H. Beck, München

English translation with revisions published 2001 by
The White Horse Press, 10 High Street, Knapwell, Cambridge CB3 8NR, UK

Translated from German by Michael P. Osman. The translation was financially
supported by the Breuninger Foundation, Stuttgart.

Set in 10 point Adobe Garamond
Printed and bound in Great Britain by Biddles Ltd

British Library Cataloguing in Publication Data
A catalogue record for this book is available from the British Library

ISBN 1-874267-47-2

Contents

Preface

Environmental history has been in existence for about 25 years, but as yet there is no agreement on its subject matter or methodological foundations. It originated as an historical subdiscipline in an attempt to identify forerunners of present environmental problems, search for causal agents and find mental or structural alternatives to a destructive relationship to nature. To that extent environmental history is an intellectual offspring of the conservation movement, which sprang into being in the late 1960s in most Western countries.

A whole series of scientific enterprises responded to the vision of an imminent environmental crisis. Today, in most cases, research on environmental issues has moved away from these heroic origins. The environment is no longer exclusively a 'Green' issue situated on society's radical margins but has become a normal topic in science and politics. This professionalised handling of environmental issues has led to a certain sobriety on the one hand, but on the other it has sharpened the methodological instruments. Today it is not enough to show a friendly attitude towards nature; the conceptual framework of environmental science must also become clearer.

Environmental history shares with history in general a disinclination for discussing fundamental methodological and conceptual principles. Of course, there are many hidden assumptions about what is important, what is accepted as an explanation and what is wrong or irrelevant. There are common guesses about the timeframe of social relations to nature, i.e. more or less explicit models of periodisation. Finally, each environmental historian has his own ideas about what is meant by the terms 'environment', 'nature' and 'ecology', the relationship of which to social and economic processes is the subject of inquiry.

It is inevitable that environmental history will develop an interdisciplinary scope, since it deals with the dynamics of natural and social processes. It touches upon historical geography, human ecology, historical anthropology and sciences such as zoology, botany, geology etc. An environmental history that remains within the boundaries of proper historical methodology, i.e. concentrates on written sources, must restrict its endeavours to aspects of mentality – perceptions of nature and society's adaptation to the environment. There are some cases in which the interpretation of historical sources will provide information on physical states of past environments, for example in climatic history (cf. Pfister 1999), but even here the results must be reconciled with results from scientific research on past climatic states.

Historiography reveals a long tradition of environmentalism that reaches back to ancient times. In this tradition cultural differences were seen as the results of different environmental conditions (Glacken 1967); above all, climate and soil were supposed to cause differing human ways of life. During the 19th century explanations of such differences focused on anthropological features and resulted in the fantastic race theories that flourished in the early 20th century. As a reaction to this fallacy, the social sciences of the 20th century, especially sociology and cultural anthropology, generally excluded natural factors from the explanation of social and cultural phenomena. The 'standard social science model' (Tooby and Cosmides 1992), which finally gained acceptance, aimed at an endogenous explanation of cultural phenomena. These were supposed to be the results of autonomous communication processes with no relationship to natural conditions. Thus, nature as a potentially extra-cultural factor could become a mere social construct.

However, an undercurrent remained in social anthropology that aimed at decoding cultural phenomena as adaptations to specific environmental conditions. For some environmental historians the works of scholars like Julian Steward (1955), Leslie White (1959) and Marvin Harris (1978, 1979) provided a first paradigmatic approach to ways of thinking about the relationship of human societies to their natural environment. A further impetus to transgressing the boundaries of dominant social constructivism was found in epidemiological history. Alfred Crosby (1972, 1986), Emmanuel Le Roy Ladurie (1973) and William McNeill (1976) reconstructed the historical relationship between parasitic micro-organisms and human populations, demonstrating the existence of a secret ecological history. These works made it clear that historical processes have a foundation in natural processes, which human agents cannot eliminate at will, and that they play a role in shaping events and moulding structures.

A further important, if not fundamental, natural dimension of historical processes is the availability of energy resources, and that is the subject of this book. Energy flows are basic features of ecological systems. It is energy that propels all material processes. When the energy systems of the past have been reconstructed, we will understand the natural framework that determines the physical boundaries of economic development. Energy issues were an important focus in the environmentalist debates of the 1970s and 1980s. The controversy over nuclear and solar energy as alternatives, and the distinction between hard ix

and soft energy paths (Lovins 1977), were based on the assumption that the management of energy flows was vital in environmental integration.

When I began to deal with the history of energy from a global history perspective some 20 years ago, it was within the context of this controversy. In 1982 an earlier version of this book was published in Germany. In those days I tried to assess the truth of whether the character of an energy system determines future paths of social evolution – as was claimed by the opponents of nuclear energy. In this context I studied the historical transition from the agrarian solar energy regime to the use of fossil energy, which fuelled the industrial transformation of the last 200 years. I came to the conclusion that the analysis of historical energy systems provides a fundamental pattern of explanation that contributes to an understanding of the basic structure of different social formations. The availability of free energy defines the framework within which socio-metabolic processes can take place.

The energy explanation of historical processes has a history of its own (cf. Martinez-Alier 1987). Energy as a physical concept only originated in the early 19th century when the transformation of thermal to mechanical energy was no longer simply technologically feasible but could also be explained theoretically. Scientific developments in thermodynamics, physiology and (agricultural) chemistry provoked first sketches of an energetic-naturalistic theory of society from which lines of influence can be drawn to more recent work, such as Eugene Odum's (1975) human ecology or Roy Rappaport's (1968) cultural ecology. These theories of energy were first articulated by natural scientists like Eduard Sacher, Leopold Pfaundler or Bernard Brunhes but provoked vehement disapproval from the social sciences, which were then in the process of establishing themselves. One major example of this resistance is Max Weber's (1909) extremely negative reaction to the 'energetic theory of culture' developed by the chemist Wilhelm Ostwald (1909, 1912). This opposition must be understood as an endeavour of the incipient social sciences to establish a peculiar field of research in which its own principles of explanation were polemically opposed to the natural sciences.

Ostwald's fate was shared by the British chemist Frederick Soddy (1921), who tried to construct a physical foundation for economics. In his naturalistic perspective the concept of economic growth was merely a convention without physical meaning. He saw the economic process as an accumulation not of material wealth but of credit or money – a mere game of fictitious financial operations. Contemporary economists did not receive this theory with much favour, but some of his ideas were later adopted by Nicholas Georgescu-Roegen (1971) and became part of the theoretical foundations of ecological economics.

In cultural anthropology it was mainly Leslie White (1943, 1949, 1954), who used energy-based approaches for the explanation of cultural phenomena. His work influenced the later development of ecological anthropology. In current anthropology, the energy explanation was mainly used by Richard N. Adams (1975) in a non-metaphorical mode. In sociology, an important and fundamental conceptual study was Fred Cottrell's *Energy and Society* (1955), a seminal work that unfortunately is almost forgotten today.

For a long time historians have either only dealt sporadically with energy problems or have treated them within the context of technological and economic history (e.g. Nef 1932, 1934/35, 1957). This changed in the context of the energy controversies of the 1970s. Energy issues now achieved much publicity and historians were asked if energy had any meaning in the past. In 1971 Scientific American published a special volume on *Energy and Power* with articles on forms of energy use in different societies. Since then several papers have been written, of which I will only mention some of those in English: Dyer 1976, Nef 1977, Ryan 1979, Thomas 1980, Melosi 1982. In addition, two major works must be mentioned: in 1988 Edward Wrigley supplemented his concept of an historical transition from organic to mineral resources in the course of industrialisation with the energy aspect, thereby proclaiming the importance of energy in the industrial revolution. The historian of technology Vaclav Smil (1991, 1994) made available large quantitiess of material, statistics and physical data for an historical analysis of energy.

This book is mainly an essay on environmental history that focuses on energy as the physical framework of material activities. Its empirical foundations are a study of the question of wood scarcity in the 18th century. My research on this topic was carried out in the early 1980s. Since then much empirical research on German forest history has taken place and we now have valuable regional studies available (e.g. Allmann 1989; Selter 1995). In addition, there has been research on the question of fuel shortage and its management in the wake of the industrial revolution (Marek 1994; Grabas 1995; Harnisch 1997; Radkau 1997; Sieglerschmidt 1999). Not all of these new results could be included in this book but I believe that my general argument is still valid, and I am certain that this text contains no conclusions that cannot be sustained.

Last but not least I have to thank two persons without whom the new English edition of this book would have been impossible. Verena Winiwarter persuaded me to re-edit the text and Helga Breuninger of the Breuninger Foundation funded the costs of the translation.

Rolf Peter Sieferle
Heidelberg, Fall 2000

I

Energy Systems and Social Evolution

The earth's biosphere is a powerful solar energy system. The innumerable living organisms that have developed into an intricately structured, all-embracing system during billions of years of organic evolution can only exist because of a steady flow of energy. A living creature is an extremely 'improbable' material entity; its decay and approach to thermodynamic equilibrium can only be prevented or delayed by a supply of high quality energy. Living organisms are islands of order in a chaotic world. As open systems they depend on the export of waste heat to their environment and can only sustain their high state of intrinsic order by importing high quality energy (so-called free energy) from their environment and releasing diffuse energy or heat to a higher level of entropy.

During the course of their evolution, living creatures have found two major ways to tap into such flows of metabolic energy. Green plants use solar energy directly by building up their biomass using a highly complex photosynthetic process in which they transform radiation energy into chemical energy. In turn, they form the energy base of herbivores, which use the chemical energy of plant biomass to construct and maintain themselves. Therefore, all animals are energy parasites of the vegetable kingdom and in the end their sole energy source is the sun. The metabolic process in which all living creatures exchange inorganic substances with their environment is mediated by an irreversible flow of energy. In essence, every process of life is basically an energy process.

Living creatures take free energy from their environment, which in turn permits them to perform work and release low quality energy. They do not 'consume' energy, since energy in the physical sense cannot be used up, but merely transformed. However, in the process it loses its usable form and becomes unusable, i. e. it is only the ability to perform work that is 'consumed'. In this way the potential for order in highly organised energy is transformed into the complex structures of life, while in the end less ordered energy barely capable of work is released to the environment as waste heat. This process of irreversible energy transformation is the foundation of all life, including that of humans. Not only the construction but also the maintenance of living states of order relies on successfully sustaining this energy flow. If the stream of energy dries up, living structures collapse and dissolve into an unordered equilibrium.

Vital to every organism and every species is the extent and certainty with which it manages to tap into the energy flow of the biosphere in the course of evolution. The ecological web of species forms a functional and metabolic system that was created through co-evolution and exists in a state of permanent flux. Every individual must tap as successfully as possible into the energy that flows through the system and is largely available in the form of other creatures that are preyed on for food. Not only must individuals preserve themselves by seeking nourishment and protecting themselves against enemies and competitors, they must also produce and frequently provide for offspring. Therefore, animal life at least requires free energy, not only to maintain its own structures metabolically, but also to perform mechanical work by converting energy into a physically useful and suitable form. This applies to all animals including human beings.

In the end, in a natural ecosystem only as much energy is available in a given area as is provided by solar radiation and can be stored. The amount stored, that is, the energy available per unit of area and time, depends on the efficiency of storage. This can vary to a certain degree and is open to innovation. If the availability of water and minerals is unlimited, plants are an unparalleled energy store. This is particularly true if one considers their entire life cycle. Plants possess no other source of energy and in evolutionary terms tend towards working with a positive energy harvest factor. The structural parts that enable them to connect positively to the biosphere's energy flow consist of the same biomass in which this energy is stored. This fundamentally differentiates them from solar collectors.

An annual average solar radiation of c. 4–8×10^6 kJ arrives on a square metre of the earth's surface, depending on latitude, clouding etc. About 1–5% can actually be fixed and stored photosynthetically by plants. The vegetative biomass of autotrophic organisms is the energy source for all animal or heterotrophic life. Plant eaters (herbivores) can only use about 10–20% of the energy stored by their food and carnivores 10–20% of the energy in herbivores. That is to say, the available energy in food declines by a factor of ten on every trophic level, which means that the need for space increases by an order of magnitude. Of the radiation energy of c. 4.2 million kJ, which falls in the course of a year on a square metre in Europe, a mere 800 kJ remain for a carnivore such as a meat-eating human (Odum and Reichholf 1980, 64). It is evident that with a given energy utilisation per capita the potential population density depends on its trophic level, that is on whether it is a herbivore, a carnivore or a higher level carnivore, which in turn feeds on meat eaters. With each step up on the food pyramid, the possible population density per unit area declines by a factor of five to ten.

Like most primates living today, the phylogenetic precursors of humans were omnivores eating a large proportion of plant foods. Omnivores not only eat a varied diet, but it is actually essential they do so to avoid deficiencies. Their population density is of necessity lower than that of pure herbivores because they cannot digest the most commonly available plant food, cellulose, which occurs in grasses. It is assumed that in the course of hominid evolution, nutritional preferences shifted towards more meat. Most likely in response to climatic and ecological changes in their environment, they were dependent on increasing their territory and, therefore, decreasing population density.

Increased meat eating brought significant advantages. Since hominids, unlike other predators, did not lose the ability to digest plant foods, they always had the ability to return to them if prey animals became scarce. Anyway, they could never entirely do without plant foods. The human organism is, for example, incapable of synthesising ascorbic acid (vitamin C) and must take it in with its food. In marginal situations this could lead to problems. Thus, the Inuit of Greenland were forced to consume the stomach contents of herbivores in winter to avoid becoming victims of the deficiency disease scurvy.

1. Palaeolithic Hunter-gatherer Societies

During the long duration of the Pleistocene (from c. 2 million to 12,000 years before present) first a process of organic evolution and then one of extra-somatic cultural evolution took place in hominids. The social and metabolic (to avoid the term 'economic') basis of these groups was hunting and gathering. They attempted to provide their subsistence by hunting but it can be assumed that they depended in caloric terms on plant foods. Like all living creatures they had to consider how to minimise energy use while obtaining their food, in other words the work put into the search for food should stand in as favourable as possible a relationship to yield. Since they did not control their resource base, the possibilities of improving energy efficiency remained within narrow limits.

Little is known about the conditions of big game hunting as it existed in Eurasia at the end of the last Ice Age approximately 12,000 years ago. From an energy perspective, using energy yield per hour of work as the measure, it may be assumed that it was more efficient than other forms of hunting (cf. Boyden 1987). Apparently even then a contradiction between 'ecological' and 'economic' efficiency could arise. The ecological energy efficiency of feeding on plants is higher than for a pure animal diet. On the other hand, the economic energy yield, construed as the relationship of work input to yield output, is higher in big game hunting than in gathering plants. The hunters were not 'ecologically' oriented in a strict sense, since they did not align their behaviour with the energy flow of the entire system, but were capable of 'waste'.

In contrast to the position in later agriculture, technological progress in the sense of improvements of energy efficiency played virtually no role in hunting large game. Thus, a metal spear tip would barely increase hunting success, but an iron sickle was clearly more effective than a flint sickle. In any case, the energy efficiency of big game hunting was far greater than in agriculture, at least as long as there was big game to hunt. Since the nutritional basis was not actively controlled (as in agriculture) but only readily available resources were skimmed off, technological progress would lead merely to quicker exhaustion of a given stock. This may be the reason why there were few innovations over long periods of time: innovations could in effect be self-destructive and, therefore, paid no evolutionary dividend.

In these circumstances, there was an optimal size for both territory and group in hunter-gatherer societies, which depended on the natural stock of plants and animals. If group size and environmental conditions (climate, vegetation and fauna) remained constant, it was possible for such a society to live in a stationary equilibrium: it skimmed off just as much energy as its ecological niche could permanently supply. If the chemical cycles remained reasonably closed, hominids could live in a given habitat as long as it existed. There was no necessity for a dynamic to exist that would lead to progressive disequilibrium.

Nevertheless, fluctuations always occur in nature, and no ecosystem is stable over the long run. Not only can the 'equilibrium' of an ecosystem be understood as the average of greater or lesser oscillations, but there are also strong fluctuations, incursions of external factors, evolutionary thrusts and collapses of ecosystems. The causes for these may be purely exogenous – climatic changes, impacts, volcanic eruptions, changes in the earth's crust – but they can also be endogenous, i. e. emanating from living creatures. New species may invade, old species may become extinct or change their behaviour. All this leads to major or minor structural shifts in ecosystems through which participants put each other under selective pressure.

With the evolution of human beings a new dynamic element entered the process. Their behaviour was not so much grounded in genetics as controlled primarily by cultural mechanisms. This meant it could change within time periods that were extremely short in biological terms. This extra-somatic evolution, which was no longer tied to DNA molecules as information carriers, permitted an enormous acceleration in the formation of ecologically relevant characteristics. The process of change by leaps and bounds that was associated with both the Neolithic and Industrial Revolutions would not have been possible on a purely somatic basis, given the sluggishness of organic evolution.

Much speaks for the view that in functional terms cultural evolution is equivalent to organic evolution (Sieferle 1997). Since the preconditions for cultural evolution were created in organic evolution, cultural mechanisms must initially have been favoured according to the same criteria as genetically and physiologically created traits. Originally, it should not have been possible to differentiate the adaptive effects of organic and cultural evolution, but their pace of development would have been dramatically different. In any case, with cultural evolution a factor came into play that could lead to a comprehensive and rapid transformation of ecological conditions.

The essential objective of any population is the regulation of its size in relation to the territory from which it draws the basis of its subsistence. On a superficial level this adaptation occurs through the availability of the sustenance in which each species specialises. It is particularly important for higher level animals to develop strategies so that they do not permanently live on the verge of famine. If a group of predators succeeds in stabilising its population size below the carrying capacity of a territory, it is superior in capturing prey to a competing population that lives on the margin of subsistence. Therefore, it can eliminate the latter, and there is an evolutionary premium on optimising and not maximising offspring. Consequently, many species of animals are capable of keeping their population numbers below the theoretical maximum. This may occur through defence of a territory, decimating superfluous young, excluding a part of the population from reproduction and even intraspecies aggression due to stress phenomena that arise with overpopulation.

Hunter-gatherer societies also possessed behavioural modes that caused their population size to remain under the maximum determined by food availability. Here, natural limitations due to mortality and a birth rate lower than the biologically possible may be cited. Often intervals between births were four to five years, which was in part due to ovulation disorders caused by physical exertion and frequently also to increased breast-feeding periods. Mothers who carried their children with them during migration and nursed them to an age of four years had only a minimal capability to conceive. An increase in the lactation period was frequently also demanded for cultural reasons. Furthermore, a multitude of direct forms of culturally mediated population control existed (Harris and Ross 1987). As a result of this behaviour, primitive societies apparently did not live permanently on the verge of famine. Individuals who survived the threshold of birth and early childhood, were quite well off (Sahlins 1974).

Insofar as one can draw conclusions about primitive societies from the study of present hunter-gatherer societies, they were well nourished as regards the quality of their nutrition (protein, vitamins, minerals) but were rather close to the scarcity limit in terms of calories. The usual intake among modern hunter societies is 8,000 kJ per day and person, which is higher than the level in traditional agricultural societies but far below that of industrial countries. Energy-rich nutrients, especially carbohydrates and animal fats, are scarce. Muscle meat of large animals in Africa has a fat content of only 4% as compared to modern farm animals with 25 to 30%. However, the hunting efficiency of modern hunters such as the !Kung San, from whose nutritional habits these values were derived (Lee 1968, 1979), is low due to the scarcity of large game in the Kalahari desert. The caloric efficiency of Palaeolithic hunter-gatherer societies may have been higher. The following table provides a rough overview.

Big game hunting (favourable environment)	40,000–60,000
Big game hunting (poor environment)	1,000–25,000
Nut collecting	20,000–25,000
Small game hunting	4,000–6,000
Foraging	3,000–4,500
Shell collecting	4,000–8,000
Peasant agriculture	12,000–20,000

Table 1. The Energy Yield of Various Methods of Production
(Kilojoules per hour of work. After Cohen 1989

It is evident from this table that hunting big game with a primitive spear was more efficient than hunting small game with the advanced technology of bow and arrow. The energy yield in the transition from big game to small game hunting (small mammals, birds, fish) clearly dropped initially, but rose again with the transition to agriculture. From the perspective of universal history, this means that the Neolithic revolution, i. e. the development of agriculture after the end of the last Ice Age, was an improvement in energy efficiency compared to the preceding stage of small game hunting, but not compared to Palaeolithic big game hunting.

Carnivores have a greater demand for territory than herbivores. The lower the trophic level of an organism, the smaller the area from which it draws its food can be. Societies of hunters-gatherers usually are, and need to be, geographically mobile, because they are at the top of the food chain and therefore require a large living area. This can be clarified through the following table:

Bushmen in the Kalahari	0.1
Aborigines in Australia	0.05–0.1
Prairie Indians in North America	0.04–0.05
Paleolithic France	0.02

Table 2. Population Density of Various Hunter and Gatherer Societies

(Persons/km^2. After Boyden 1987)

The mobility of these societies is dictated by their lifestyle. Sedantarism is only exceptionally found among hunter-gatherers, for example if they succeed in tapping into a stream of resources such as a migrating herd or fish. However, mobility solves a number of problems: if a group becomes too large it divides; if there are conflicts one leaves the group. The same applies to handling environmental problems: if the camp is too dirty, if garbage and faeces begin to smell, if the huts are full of vermin, if the waterhole is polluted, if illness or unexplained deaths occur, then this inhospitable place is left and people move on.

Migration soon required adaptation to an environment that did not suit the natural organic equipment of humans. They had to develop 'technical' skills to adjust to new conditions. An important step was the mastery of fire. Fire permitted the settlement of northern territories to which humans were not adapted by nature. With its help they could conquer new habitats after the old territories were fully occupied under given technological standards and food preferences.

With systematic mastery of fire, a first step was taken in the technological use of energy (Goudsblom 1992). The forms of energy utilisation before the use of fire were still closely tied to the organic processes of metabolism and biological energy flow. Fire created a microclimate in artificial spaces that resembled the one to which humans had adapted organically. It permitted them to live in areas into which they could only evolve after very long-term evolutionary processes resulting in the reacquisition of fur, an increase in body volume to reduce the relative size of surface or the development of fat deposits under the skin. Furthermore, humans could expand their range of food: roasting, grilling and cooking of meals enabled them to eat foods that were hard to digest uncooked. Many plant species contain chemical substances that are poisonous or cause nausea as a defence against being eaten. Cooking can in part denature these substances so that the plant becomes edible. Other plant substances are made available by fermentation, while smoking, drying or repeated heating can preserve foods.

Difficulties stood in the way of cooking in the strict sense. Watertight and fireproof vessels were required for cooking and steaming. Hollow stones but more particularly fired clay were available. The mobility of the hunters' existence did not permit them to load up with many heavy dishes. Although hardening of clay had been known for a long time – it is an experience one often makes while setting up a campfire – non-sedentary groups used virtually no earthen vessels, but only buckets and pots made of leather. They could make soup by tossing hot stones into the broth, but the food was not as well cooked as over an open fire. But it was only worthwhile to produce fireproof, fired pots in a sedentary state, and it was only then that those plants became edible that required a higher temperature. The range of foods was enlarged and population density could rise.

Except in their use of fire, primitive hunter-gatherer populations were hardly different from predator populations in terms of energy usage. However, they were ubiquitous and did not specialise in particular types of food. They did not eat everything that was edible in a territory, but the range of their food was highly diversified compared to that of other animals. There was a strong gradient of preference in nutrition: certain plants and animals were only eaten if nothing else could be found; on the other hand they went on long trips to capture particularly valued prey. Hand axe, spear and eventually bow and arrow were technological aids in obtaining food. By using a series of mechanical principles (lever, wedge, inclined plane) results were achieved that would have been unattainable to humans without tools. In all of these areas of work, humans remained the only energy converters. Their own bodies transformed the chemical energy of ingested food into the mechanical energy of muscle power. They developed technical means to focus labour to achieve an improved result; but the mechanical aids were not different in principle from the equivalent organic tools of animals. However, humans were the only ones to possess so many tools at once: they were generalists in every respect, settled all climate zones of the planet, used all food reserves and employed all the mechanical principles available.

The energy system of Palaeolithic society was characterised by humans tapping ever more efficiently into energy flows. But in contrast to agricultural society they did not reconfigure these energy flows. Therefore, it is possible to speak of a regime that utilised unmodified and uncontrolled solar energy. Humans skimmed off free energy wherever they found it but employed almost no buffers, nor did they attempt to control energy conversion actively and permanently.

2. The Neolithic Revolution and the Problem of Dynamics

Approximately 12,000 years ago the secular transition to agriculture began. This so-called 'Neolithic revolution' probably constituted the most profound leap in the history of humanity. Only the Industrial Revolution has a similar fundamental importance for life-style, ecological embeddedness of humans and their relationship to other species. It is not possible to deal in detail with the broad anthropological discussion of this transition here. It arose after the publication of the theses of Ester Boserup (1965) on the origins of agriculture during the Neolithic, according to which there was a close relationship between the population growth of Neolithic hunter-gatherer societies and the transition to agriculture.

This position turned forcefully against an older school that was strongly indebted to 19th century social Darwinism and the idea of progress and sought the 'invention' of agriculture in an autonomous improvement of technological knowledge. The fact that hunter-gatherer societies possessed all the components of the knowledge that agriculturists employed speaks against this view in practice. Also, there are numerous documented examples of game hunters living next to farmers without seeing the necessity of assuming their 'progressive' life-style.

The explanation of the Neolithic revolution as an innovative reaction to autonomous population growth, as it has been presented in extreme fashion by Cohen (1977), encounters a difficulty. If population grew steadily during the entire Pleistocene, then the rate of growth must have been exceedingly low, so that it is incomprehensible how overpopulation could have arisen within such a short time period. If we assume a hypothetical growth rate of 0.0005% annually, we obtain a growth factor of 148.4 over one million years. This is quite high since it is assumed that population at the close of the Pleistocene was about 6 million, so that the initial population would have only comprised 40,000 persons. Then, for a period of 10,000 years the growth factor is only 1.01, which hardly could have led to overpopulation.

Estimates and interpolations of growth rates of this kind are probably an anachronism stemming from an age that experienced continuous growth and projected it back into the past as a matter of course. As soon as two (usually estimated) figures are available for comparison, growth rates are calculated. This is a particularly favourite game of historical demography. Livi-Bacci (1997, 31) is certain that 252 million people lived around the time of Christ's birth while only 6 million lived 10,000 years before. Therefore, 'annual growth' was 0.037%. Why? Simple: 252/6 = 42. Of this the 1/10000th root = 1.00037. Voilà!

Closer consideration demonstrates that talk of 'growth' is only reasonable if we are faced with steady processes; but this is highly improbable if we are dealing with long periods of history. The assumption that there could have been long-term continuous growth of any physical parameter is simply aberrant. Simple calculations will demonstrate this. Let us assume that a population of 6 million did grow by 0.037% annually. This would mean that 2,220 persons would have been added in the first year. In the second year the population increase would have been 2,220.82 persons, in other words the additional increase would have been less than one person.

Or another calculation: let us assume that the original population of 6 million hunter-gatherers was living in groups of 50–60 persons, so that there would have been about 100,000 groups globally. If the entire population had grown by 2,220 persons, an individual group would have increased on average by 0.0222 persons, i. e. each group could only grow by one person every 45 years. That would mean that growth was not perceivable and the calculated trend lay well below the volatility of population units.

So we have the alternatives of a growth rate that was so extremely low that it was imperceptible as a controlling behaviour variable in the relevant population, or if it was noticeable at all, the population quickly reached astronomical proportions. The annual growth rate of the population of Europe during industrialisation was barely 1%. If we interpolate this rate for 10,000 years, a growth factor of more than 10^{43} results, i. e. beyond what can be imagined! However, this means that the occurrence of growth with a noticeable rate is something unusual in historical terms and requires explanation, and not the other way around. Phases of growth are under suspicion of being historically exceptional from the start.

These considerations preclude the idea that 'natural population growth' suddenly passed a critical threshold 12,000 years ago and provoked the transition to agriculture. If there was growth-induced population pressure at the end of the Pleistocene, it must have been because there was a drastic increase in population in a short period of time, a rupture of an older equilibrium. Then an explanation of why this was the case becomes necessary.

It is remarkable that the transition to agriculture apparently occurred in different areas remote from each other, and not only in Eurasia but also in America (Smith 1995). Early agricultural regions were islands in territories inhabited by hunter-gatherers. The argument that agriculture spread through diffusion, due either to migration or to transfer of knowledge, is hardly credible. The only explanation that does not need diffusionist models is climatological (see, for example, Childe 1942). Only climate acts universally and is not mediated by human intercourse and, in fact, the Neolithic revolution fell within a period of global warming. However, there were frequent climatic fluctuations and interglacials in prehistory; the argument is therefore not sufficient to explain why the transition to agriculture should have occurred at the end of the last Ice Age.

With such a complex process as the Neolithic revolution (and later the Industrial Revolution) it is probably more appropriate to look for a non-causal, i. e. a functional explanation. Surely there is no single determining causal line, but a number of factors must have coincided to effect the decisive shift or breakthrough. In processes of universal history it is probably reasonable to differentiate in a model between the normal course of events and specific discontinuities during which a technical and economical, socio-political and cultural-ideological regime is restructured. During the restructuring phase, which is simultaneously the destruction of a previously existing system, components are regrouped, take on a new form, produce subsystems with emerging characteristics and form an altogether new structure.

In the initiation of a process of this kind, a *convergence* of different factors will play a critical role. It need not be assumed that the collapse of the old regime is already predetermined during the contingent appearance of certain novel elements and that the structure of the subsequent regime is already completely fixed in it. Instead, it seems reasonable to assume that we are dealing with an open 'bifurcation' if a particular threshold of destabilisation has been reached. Then the future course of the development can only be stated with a high degree of uncertainty. In contrast to periods of stability, when it can be said with some certainty what future developments may be anticipated – even though prediction with much precision may hardly be possible due to the great complexity of social systems – this is impossible in principle in a bifurcation. At best, one can preclude certain developments as extremely improbable and consider others more likely. For purposes of historical reconstruction, it is critical if a more probable field of development is favoured by factors that formerly appeared isolated in all sorts of contexts without being able to effect anything very much. It is only the meeting of such factors that can create a destabilising situation with relatively open bifurcations.

According to this model a certain line of development cannot be reduced to one or more causative 'reasons' but only to a particular constellation of factors. If we assume coherence in an overall societal system, then fluctuation in one subsystem must provoke processes of adaptation in other subsystems. These generally consist of a buffering of fluctuation: the functional failure or modification that is associated with fluctuation is compensated in other areas. The overall system changes to a certain degree but neither its environmental profile nor its stability and ability to reproduce need suffer. It maintains its identity. Such buffering may be anticipated when relatively unimportant fluctuations occur or when no strategically important aspects are affected.

On the other hand, oscillations may be enhanced in a virtually self-promoting manner. In this case, a developmental thrust may occur in which the entire order is structurally rearranged. For this to happen there must be a readiness in individual subsystems not to suppress the innovation, but to enhance it and adapt to it. This is particularly so if a favourable instability already exists for other reasons and only requires an external nudge to cause the system to topple. In that case it is possible to speak of bifurcation.

A transformation of the entire system, i.e. the historical inauguration of a new social metabolic regime, can hardly emanate from one isolated subsystem. It may be assumed that isolated fluctuations are generally compensated. In a sense, an 'overdetermination' is required for such fundamental historical transformations as the Neolithic and Industrial Revolutions. The technical and economic, socio-political and cultural-ideological subsystems of a society must be internally coherent. Furthermore, they must match functionally in such a way that the system can maintain its boundary to the external environment. Therefore, if a readiness to vary in one and the same direction occurs in all three areas, an accelerated, self-enhancing development can set in. However, it remains crucial whether positive feedback loops build and whether the innovation is strengthened by them. Isolated occurrences of events usually remain irrelevant: this is demonstrated by the inconsequential existence of so many 'precursors' of historical phenomena, to which great significance is attached when a dynamic transformation is explained. Therefore, it is idle to ask for the 'cause' of a system-transforming development, since every aspect of it may have occurred by itself at some point without a comparable effect resulting. Highly complex phenomena such as the Neolithic and Industrial Revolutions can be explained with a model of this kind, but of course this makes empirical validation of the model extraordinarily difficult.

During the Neolithic revolution climatic change (global warming), population growth and a new social-metabolic regime, agriculture, coincided. What processes in the socio-political and cultural-ideological regime were related to it can only be reconstructed with great difficulty, if at all. Judging by the result, it can be said that property and power relationships took the place of kinship relationships as the central form of organising social cohesion. Proprietorship of land fits functionally with agriculture; kinship relationships are by contrast the simplest and most plausible method to create social cohesion under the conditions of hunting and gathering (White 1959). Since human population density is mainly controlled by cultural rules, an autonomous variation in the cultural area may have been a 'causative' factor. But, of course, this cannot be determined retrospectively.

In any case, it becomes apparent that the complex agriculture/high population density/property is superior in evolutionary terms to the complex hunting and gathering/low population density/kinship relations in the sense that the former has established itself irreversibly against the latter. This is not as obvious as it may seem at first sight.

If the food supply of hunter-gatherer societies is compared to that of agricultural societies, the latter does not appear as favourable as one might expect. Hunter-gatherers are as well provided with calories as agricultural societies and their nutritional range is broader. They therefore suffer less from lack of vitamins and other deficiency diseases. Qualitatively, the nutrition of hunter societies is better and more balanced than the monotonous carbohydrate diet of agriculturists. Agriculture limits the number of prey animals and useful plants considerably, and places an unusually strong emphasis on the consumption of grasses in the form of cereals, maize, rice and millet. Agriculturalists thus become strongly dependent on the vegetative cycle of these preferred plants while the nutritional basis of hunter-gatherers is much better buffered against gaps because of its greater breadth. Famines must have been more frequent in agricultural societies that were affected by harvest failures than in hunter-gatherer societies, unless these had been squeezed into entirely marginal areas.

In general the implications of the Neolithic revolution in nutritional physiology can be summarised as follows (cf. Cohen 1989). There was

i. Herbivorisation, that is a decline in the share of meat in the diet;

ii. Cerealisation, that is a decline in the proportion of fresh to stored, preserved and processed foods, with grass seeds taking centre stage;

iii. Standardisation, a reduction in the variety of food; and

iv. Quality reduction, a reduction in the quality of the diet in the sense that a deficiency of proteins, vitamins and trace elements could occur.

There is evidence then that the material 'standard of living' declined during the transition to agriculture. Humans were nourished less well, but at the same time they had to work more and especially more continuously. The life-style of hunter-gatherers relied on tapping passively into a flow of energy. Since this flow fluctuated by nature, they had to maintain a safety margin to be able to survive a bottleneck. This meant that usually they were living in superabundance, which was expressed in a low requirement for time to obtain food.

Agricultural societies, by contrast, controlled the flow of resources much more closely and, due to their sedentary state, were able and also forced to store foods. Because of storage they could count on a far steadier supply of food with the consequence that they could reduce the safety margin for a possible situation of minimum supply. This had far-reaching consequences: it not only meant that they used their natural environment more intensively, but also that the investment in labour rose considerably. Peasants, unlike hunters, could eat bread – but they had to do it in the sweat of their brow (see Müller-Herold and Sieferle 1997).

The increased physical burden on farmers from heavy work and a relatively poor, low-protein diet are confirmed by prehistoric skeletal finds. In Greece and Turkey average heights of 1.78 m (men) and 1.68 m (women) were found among postglacial hunters. In agricultural societies that settled the same area around 4,000 BC the heights were only 1.60 m and 1.54 m. At the same time there were signs of a social differentiation in the diet: members of the upper class were larger than peasants.

After all this there should be no doubt that the transition to agriculture was not simply an 'improvement of the human condition', but was linked to a worsening for the majority of the population. The only real evolutionary advantage of agriculture was that more people could live in a given area with this mode of production. Agriculture is not easier, nutrition is not better and supply is not more secure than in hunting and gathering. Its main effect lies in bundling the energy flow with a dramatic consequence: it makes more food energy available per unit of soil and time, and this means that the population density can rise dramatically (Cohen 1977, 15).

3. Traditional Agriculture – A Controlled Solar Energy System

It makes no difference to the human metabolism if the animals and plants eaten are wild or if they are bred and cared for. Also, the effort of procuring a certain amount of calories from a preferred food in a natural ecosystem is initially almost as large as but not larger than in agriculture. In the course of development agriculture permitted controlled formation of usable biomass through photosynthesis and animals, so that a maximum yield of energy could be gained with a minimum investment of work. Agriculture is in this respect a solar energy system controlled by humans: solar energy is stored photosynthetically by plants that are selected, bred and cultivated in such a way by humans that a large part of their biomass can be monopolised for their purposes. This occurs in a multiplicity of ways, depending on local circumstances and technical skills. In principle agriculture means nothing more than selecting particular plants as crops from the diverse range of species that inhabit a natural habitat, to ideally live alone in their habitat, the arable. For this purpose their competitors for light, water and soil nutrients will be combated and removed as 'weeds'; also, they are protected against food competitors of humans, 'pests' and 'vermin' of all sorts.

A field regularly cultivated and harvested by humans is from an ecological point of view in a very early stage of succession in which net biomass production is very high. This means that more biomass is formed photosynthetically than is consumed by respiration at the same time. The creation of a monoculture of the preferred plant by destroying other plants and their seeds through burning, hoeing and ploughing initially serves this purpose. Thus, a larger amount of human food can be produced in a specific area than if one were to find food plants in a natural plant society. Another fundamental agricultural strategy consists of enlarging through breeding those parts of the plant that are supposed to serve human purposes at the expense of 'superfluous' and less desirable parts. The crops that were created in this way were symbiotically linked to their parasites – humans. These plants would be as little capable of surviving without humans as the other way around.

It is thus a characteristic of domesticated grain types that the kernels do not fall from the ears by themselves but must be threshed. Humans have bred this characteristic into these particular grasses because a larger part of the harvest arrives at the farm this way. Under natural conditions this would be a considerable disadvantage in survival: without human aid they could not spread their seeds and would soon be crowded out by wild types. The same applies to the size of seed kernels, as in maize or other grain types of which the stalks were bred down so that as much of their biomass as possible would be concentrated in the kernels. In a natural plant society their competitors would overshadow them and they would dwindle. In an environment characterised by agriculture these characteristics provide an enormous advantage in selection because humans protect them with great expenditure of labour and look after their dissemination. In this they are ahead of their competitors.

Agriculture means not only that the natural variety of the biotope utilised is eliminated in favour of a few plants, but also that the plants themselves are altered genetically by breeding. Crops are artificial bioconverters that are produced and reproduced by humans. The purpose of agriculture in terms of energy is to gain more chemical energy as biomass than the amount of chemical energy that must be transformed into mechanical energy (labour) and expended in this process. For this to happen, the ratio of energy expenditure to yield must be at least 1:5, because as energy converters humans only have an efficiency of 20%, i.e. they need a daily diet with a combustion value of 12,000 kJ to perform mechanical work equivalent to 2,400 kJ. If they cultivate a field they cannot expend more than 2,400kJ on average for an area that produces 12,000kJ within the growth period, including preparation, transport and processing. With this ratio of effort to yield the farmer can barely feed himself.

However there is great potential for improving efficiency in agriculture. Let us imagine the transition from gathering to agriculture in an idealised form: within the nutritional range of hunter-gatherer societies the harvest of wild grains may have played a role for some time. Grain may not have been a very attractive food, since it contains little protein compared to meat, nuts and wild legumes, is often hard to digest, must be prepared by a complicated process (grinding, fermenting, cooking) and may lead to deficiency diseases because of its low calcium content (rickets). But it has the advantage that it often occurs in highly concentrated form in nature, so that larger quantities may be harvested. It is particularly easily stored without requiring special conservation. During harvest of grain, care was probably taken to see that sufficient seeds reached the ground to make a new harvest possible next year. Perhaps the transition was finally made to help grain grow also in locations where its chances of survival were naturally less.

The usual grains are pioneer plants, which settle on disturbed soils but are soon displaced by other plants. These competitors must be permanently removed and fought by the farmer, perhaps with fire and the hoe. The transition from mere harvest to domestication of grain implies that the expenditure of labour rose. Farmers successively moved on to create an artificial ecological niche for their preferred crops, an 'artificial prairie' in areas in which associations of perennial plants would stabilise without human intervention. The natural environmental conditions were transformed to the extent that favoured plants could live in them and adapt in turn.

At some point in this process the critical threshold to surplus production was passed. There is much that points to the fact that agriculture owes its existence to need rather than to a wish for improvement. However, at some time the ratio of effort to yield must have improved to the point where it would in theory have been possible to reduce the effort to receive a fixed return. This did not happen. From the Neolithic revolution on, the life of peasants consisted of long workdays and much toil and labour. Many agricultural myths tell of a golden age, a Garden of Eden, from which people were driven to wrest their bread from a field overgrown with thistles and thorns. This sounds like a reminiscence of life as hunter-gatherers before the Neolithic revolution.

The improvement of productivity or technical expansion of carrying capacity of the farmer habitat was expressed by yielding a surplus. Two reactions to this are conceivable:

- With the same expenditure in labour, productivity could be increased to achieve higher gains and feed more people;

- The expenditure of labour could be lowered with constant production.

In principle agricultural societies 'selected' the former strategy even though in the context of simple peasant societies there was behaviour that could be considered in terms of the second strategy (see Groh 1992). Selection of the former strategy may have been due to the fact that there was a wish to capitalise on (military) advantages of a higher population density. In particular it may have been extraordinarily difficult to give up the former strategy once embarked on it. Growth-oriented agriculture may be an evolutionary trap, a demanding trajectory of development, from which it is impossible to disengage, comparable to economic growth in the context of industrialisation.

Four phases can be distinguished in the development of the solar energy system of preindustrial agriculture:

i.   Transition to the reproductive use of food plants and prey animals, e.g. sparing mother animals, harvesting without destruction of fruit-bearing plants, cautious and sparing handling of natural resources. This behaviour is also found among many hunter-gatherer societies, but there is evidence that hunters exterminated some species of animals.

ii. Monopolising certain prey animals and plants, by eliminating competitors for food through weeding, pest control and protection of domestic animals against predators.

iii. Concentration of species beneficial to humans, especially by establishing fields.

iv. Deliberate remodelling of species and ecosystems, first by breeding for preferred characteristics in plants and animals, then through an active change of environmental conditions. This enables plants to live in locations from which they would have been barred by nature. It involves measures such as irrigation, drainage, clearing, terracing, fertilisation etc. This finally transforms the entire landscape into a new type of landscape: the agricultural countryside.

An important element was husbandry of livestock. Its first effect was the stabilisation of food supply. Success was not guaranteed in hunting and humans were competing with predators. Keeping and breeding livestock made it possible to eliminate these competitors for food, so that a certain amount of animal produce was always available. This is particularly true if by-products were used, such as eggs, milk and wool. Animals could be bred to improve these products: cows that give more and more milk; pigs that put on much fat; hens that lay eggs almost daily; sheep that possess more and longer wool. Domestic animals would often not be capable of surviving without human protection. They must be tended and supplied with sufficient food. An attempt would be made to feed them refuse that is indigestible for humans (grass, acorns, leaves etc.). Thus by using the detour through animals humans could consume biomass that was otherwise unusable.

However, animal husbandry becomes problematic if a certain part of the land must be made available for their feed. During the transformation of plant into animal biomass the energy efficiency is about 10–20%. This means that in an area that grows food for enough livestock to feed one person, grain could be grown that would feed eight persons. Extensive consumption of livestock is only possible if either sufficient land is available to pasture animals, or the land is such that crops useful to humans, which are usually quite demanding, cannot grow on it.

Livestock-raising nomads are a special case. As far as the control of the solar energy system is concerned, they are formally positioned between hunters and farmers. A conjectural history can be outlined. Hunters permanently followed a large animal herd; the herd became used to them; they drove away other predators and took care not to kill any gestating or mother animals; they encouraged selection of certain characteristics, for example by preferentially killing aggressive male animals; in this way the herd was eventually domesticated. In contrast to cattle-raising farmers, nomads controlled their herd but not its nutritional basis. Grass grows without their intervention. If livestock-raising societies live exclusively on their herds and only obtain plant food from farmers in exchange for livestock, they must be nomadic. A herd that is large enough to feed a nomadic society requires huge pastures. Therefore, by their way of life and as far as the flow of energy is concerned, nomads are closer to hunters than to farmers. This does not mean, of course, that there is a historic succession from hunters to farmers through herdsmen, as older accounts of world history assumed, but that herdsmen are a special adaptation to unusual circumstances. They live in steppes and savannahs, where agriculture is barely possible, or in symbiosis with farmers with whom they engage in trade or from whom they demand plant food as tribute. In the latter case we are dealing with a de facto branch of agricultural production that is differentiated by a division of labour.

In the course of population growth, land pressure generally increases in agricultural societies. However, when humans and animals compete for food area, humans as a general trend displace animals. For humans, or rather the lower layers of agricultural societies, this requires them to slip to a lower trophic level. Transition to a vegetarian diet permits an increase of population density up to eight fold in a given area. People actually have to give up their preferred meat diet and become herbivores. This is the reason in terms of energy why a decrease in meat consumption can be observed in progressive agricultural societies that tend towards overpopulation. The proportion of meat in the diet is therefore a measure of land scarcity and relative overpopulation (Harris 1978).

However, in ecological terms nutritional systems cannot be viewed simply from the energy perspective. Pigs were kept despite energy losses because they supply proteins and fats that would be hard to come by on a purely vegetarian basis. Also, many animals served several purposes. For example, cattle that were used as draught animals generally ended their life cycle in the pot.

Nevertheless mature agricultural societies, like China, primarily tend towards keeping animals that demand little space: pigs and chicken that live on waste, fish and ducks that live in and on the water.

Within agricultural societies, there is a broad spectrum of cultivation types that are conducted with varying degrees of efficiency as reflected by yields. This depends not only on natural preconditions such as soil, climate, precipitation and the available plant and animal species; of far greater importance is the productivity of the farmer's labour, which may be considered as an improvement of efficiency in the energy sense.

Swidden agriculture with rice (Iban, Borneo)	850 MJ/ha
Horticulture (Papua New Guinea)	1,390 MJ/ha
Wheat (India)	11,200 MJ/ha
Maize (Mexico)	29,400 MJ/ha
Intensive arable agriculture (China)	281,000 MJ/ha

Table 3. Annual net energy yields of land under cultivation

(Including fallow. After Boyden 1987)

If the figures in this table are compared to net energy yields of hunter-gatherer societies, which amount to 0.6–6.0 MJ/ha a year, it quickly becomes apparent how sensational the differences are between these ways of life. Based on Chinese intensive agriculture, about 50,000 times as many persons can live in a given area than with hunting and gathering. This explains both why a return from an agricultural society to the life of hunter-gatherer societies is impossible, and why agricultural societies tend towards displacing competing primitive societies that have so much higher a demand for territory. From the perspective of agricultural pioneers the land of hunter-gatherer societies is massively underpopulated and unused. To European farmers, who conquered territories like North America or Australia in the modern period, these were 'empty'; to the original inhabitants, who lived as game hunters, they were actually fully settled.

4. The Structure of the Agrarian Energy System

The agrarian solar energy system depended on the use of qualitatively different energy forms that in the end can all be traced back to the sun, but which, given the state of technology, could only be converted vicariously into each other. The idea that heat, food, work and light could be summarised under the common term 'energy', or that water, wind, motion and nourishment could be the same

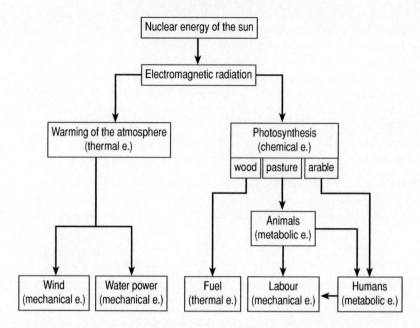

Figure 1. Structure of the agrarian solar energy system

in some way, was alien to agrarian society. Only within the framework of the industrial system did a complete conversion of different energy forms become possible, and only then could the general term 'energy' arise. However, it is possible to apply this analytical term retrospectively to past conditions.

In historical reality different types of land use were associated with the qualitative character of individual energy forms. Metabolic, mechanical and thermal energy – food, work and heat – may be differentiated. Each of these energy forms had to have a specific area dedicated from which the appropriate energy carriers were obtained: the arable was associated with the metabolic energy of human food, pasture with the mechanical energy of draught animals and woods with the thermal energy of wood. Let us consider mechanical energy in more detail.

In agrarian society the utilisation of mechanical energy gained independent significance. Since hunter-gatherer societies only used human muscle power as a source of mechanical work, a differentiation of metabolic and mechanical energy was not meaningful to them. However, in agricultural societies nonhuman biological and technical energy converters played an important role so that closer consideration makes sense.

Biological energy conversion involves the transformation of chemical energy fixed in the food biomass into mechanical energy or work with a human or an animal functioning as a bioconverter. The energy efficiency of this transformation is 15–20 %, and it makes no great difference in crude mechanical devices, such as the operation of a gin, if humans or animals are working. This can be explained with an example: The average output of a horse lies around 600–700 Watt, but that of a human around 50–100 Watt. The output of a horse is about eightfold that of a human and the relationship of the nutritional energy requirements are about the same. It is about 100 MJ a day for a horse while a physically working human needs about 12 MJ (Smil 1991).

These figures illustrate that an area capable of feeding eight humans, who would in principle be free to perform the same work, became free if the horse was given up. However, further terms must be considered in this simple calculation. First, it cannot be overlooked that the horse can only be utilised for very simple mechanical work, such as transporting loads, which on the farm is not required all the time. If the down times of the horse are considered, the estimate that the energy efficiency of a human is two and a half times as high over longer periods does not seem far fetched (Cottrell 1955, 21). Also, the horse does not look as good in comparison to other livestock: it has a relatively high demand for food, while cattle will be satisfied even over the long term with grass and hay. Therefore, it is little wonder that the horse only displaced the ox in the 18th and early 19th century, rather late, as the draught animal in agriculture. Before that it was mostly used for military purposes, where its superior qualities of mobility and learning ability mattered more than costs (Keegan 1993, 153ff).

Despite their high requirement for land, the use of work animals was unavoidable under certain climatic and ecological conditions. If the growth period is limited to a certain part of the year, work like ploughing and harvesting comes due and must be performed in a short period of time; that is, it cannot be divided up and delayed according to will. These are the reasons why the use of undemanding work animals like oxen has always been worthwhile in temperate climates such as Europe, even if they compete with humans for food-producing land. However, in certain areas of southern China, where agriculture can be practised all year round, humans displaced work animals and the plough was in turn replaced by the hoe.

An important field of application for bioconverters is the transport of heavy goods. The simplest method of overland transport is carriage. But the following problem arises here: When an item is carried – by humans themselves or by pack animals – it is not only carried in the desired direction but undergoes an up and down movement with every step. This lifting and breaking is an enormous waste of energy, so that the transport of heavy goods over long distances is hardly worthwhile. An example: a human can carry a maximum of 40 kg of grain over a distance of 25 km in one day and he consumes in this time period about a kilogram of grain. If the return trip and a day sojourn at the point of origin and the goal are counted in, he will use 16% of the carried load for a distance of 50 km, 25% for 100 km. If he were supposed to travel a distance of 500 km, he would not be able to feed himself with the grain carried along.

It is almost always an advantage if a heavy load is drawn, because then the transported item in principle only undergoes a horizontal motion, while the energy expenditure required to overcome friction is proportional. Means that overcome friction are worthwhile, such as building sleds or carrying frames or levelling paths that are often used. From this, a direct development leads from the roller to the wheel, a key innovation of technology that is set apart from the principles of organic motion.

If the wheel and wagon, which are rather costly and heavy constructions, are to be used rationally, fairly smooth and level roads must be made. It is desirable to make the wheel as narrow as possible because this reduces friction. However, this places heavy demands on the road surface. If it is too soft, a narrow wheel digs deep furrows, especially in wet conditions. Therefore, a much-travelled road must be paved with wood and/or stones, which requires heavy input of labour and was at first only customary in settlements. If no draught animals are available, road construction is often not worthwhile. In the agrarian civilisations of South and Central America, where almost the entire megafauna became extinct in the late Pleistocene, no draught animals were available, therefore no broad overland roads were made, no wagon was developed and not even the wheel was known. This was different in the agrarian civilisations of Eurasia; here in parts excellent road networks and wagons for all purposes were built. But even overland transport with horse and wagon has its energy limits. A horse eats a wagonload of fodder a week, which means that the net yield of fodder transport is already negative after a week (Ohler 1986, 141).

Under these conditions people will attempt to avoid the expensive use of draught animals and to use technical energy converters that cost less. The most important applications of mechanical converters are the use of water and wind power. The water flow which is tapped by mills originates when water is evaporated by solar heat and rains down in higher altitudes, from where it flows back to sea level. Wind is mainly generated by balancing temperature differentials in the atmosphere, except for the trade winds, which largely obtain their energy from braking the earth's rotation. This also applies to the tides that are used in special power generators. Another non-solar energy source is warm springs (geothermal energy), which are fed ultimately by radioactive processes in the earth's interior. However, useful energy is predominantly solar in origin, that is it depends physically upon nuclear fusion at a great distance from the earth.

The simplest way to use the energy of flowing water was the raft. Its use was already commonplace in Palaeolithic societies and it was used for the transportation of heavy goods, especially wood, up to the 19th and 20th centuries. Means of transportation and transported goods coincided in the raft, and there was only a very small expenditure of specific energy, which was exerted by the crew during steering. In a strict sense, flowing water was not the actual and only source of energy: the river essentially constituted merely the medium of transportation, which permitted the potential energy contained in the raft to be transformed into mechanical energy. The river as such was only used in part in terms of energy; rather it was a 'slope' with little friction, which allowed the goods to slide to the sea.

The raft had one principle disadvantage: transportation was only possible in the direction of the flow. This caused a distinctive asymmetry in land use that was widespread in the preindustrial landscape. Population and consumption centres lay towards the lower part of rivers so that they could be supplied with rafts and boats without giving rise to greater energy expenditure. Since the heavy goods transported to centres of consumption needed to be paid for, at least insofar as they were not tribute, taxes and rents, there had to be a material return flow, usually consisting of lighter (luxury) goods. For this transport against the current, labour (humans, draught animals) needed to be employed, and the creation of dedicated towpaths was necessary. Even so, towing against the current was more efficient in terms of energy than transport over land, especially if the slope was not too steep.

A significant technical converter of water energy was the water mill (Bayerl 1989). From a technical perspective it acted as a brake on the water flow and transformed its energy flow into a rotational movement that could be used for a number of commercial purposes. The water mill was originally invented by Oriental civilisations, probably from a water-raising device. Compared to the muscle-powered horse gin, it had the advantage that no specific (pasture) area needed to be dedicated to it. However, it too had problems that lay in the very nature of the availability of water. On rivers only underflow water mills, often even floating mills, could be operated, although they were not very effective, were threatened by flooding and also encumbered navigation and rafting. In addition, streams might dry up in summer and freeze in winter so that the mill would cease to function.

The major problem of the water mill lay in its being fixed to a single location. Energy made available with its aid could only be transported by mechanical transmission, i.e. with the help of bars, rollers, bands and wheels.

Since this resulted in considerable friction losses, only distances up to 1,000 m could be handled. Greater amounts of mechanical energy could hardly be made available away from rivers and streams. In particular, this was a problem for mining, which required lifting devices for draining the pits. If ore was to be mined at some distance from a usable watercourse, one was dependent on a costly horse gin.

Wind was the second important source of mechanical solar energy. Wind energy was particularly important in water transportation. The sailing ship was the real foundation of long-distance transport across greater distances. As soon as problems of steering (crossing against the wind) and navigation were solved, the metabolic effort in setting and raising sails and other work on the boat, as opposed to rowing, were minimal. Only the military sector was once again an exception. The galley could survive despite its enormous operational costs and low range until the 18th century because it did not possess the main disadvantages of the sailing ship: it was agile, it could be accelerated and decelerated and it could be used with no wind (Guilmartin 1974). The sailing ship was apparently developed and used independently in different agricultural societies. It was even employed in inland transport on river and canals where the possibilities of crossing against the wind were very limited. Its advantages compared to overland transport were so enormous that even long detours were worthwhile.

The windmill was of little importance compared to the sailing ship. It developed in the 7th century AD and only established itself very slowly. The construction of a windmill was technically less costly than that of a water mill but it had to be turned into the wind and therefore be dampened. Its main disadvantage was that its energy could not be controlled, dosed or buffered. It could not be operated with too much or too little wind, so that it could not be used for machinery that did not allow for sudden interruption, like crushing hammers or bellows for iron smelting. It therefore remained restricted to less demanding applications, such as pumping water from drained land and grinding grain.

In a level country like Denmark, ponds had to be created to obtain water power, but this had the negative side effect that groundwater rose. For better drainage of soils with the objective of avoiding acidification and waterlogging, rapid run-off of water was desirable. There was conflict between the use of water power and agricultural production that was resolved through increased use of the windmill (Kjaergaard 1994, 115).

Another element of the agrarian solar energy system was the use of chemical energy to generate both process and room heating. For this purpose plant biomass had to be burned, especially wood and peat. The use of fire was already found among early humans and was customary among all hunter-gatherer societies. Burning wood quickens the oxidation process that would occur even without fire through metabolisation by small animals and microbes. The utilisation of wood by agrarian societies was in principle nothing new, but it occurred on a completely different temporal and spatial scale, and led to an extensive encroachment on woods.

The production of plant biomass for the purpose of generating heat required permanent use of a particular area for firewood. Resorting to a larger energy stock, the primeval forest, was only possible in pioneer stages such as the conquest of America. However, in the permanent utilisation of agrarian space the consumption of wood needed to be synchronised with biomass production. Much more will follow below.

Three different types of land can be attributed to metabolic, mechanical and thermal energy in agricultural space: arable, pasture and woods. In agrarian society food, heat and motion had a qualitative character because technically useful conversion of one energy form into another was impossible. If, for example, sufficient quantities of mechanical energy were available in the form of a watercourse, there was still no possibility of transforming it into useful heat. On the other hand, plant biomass could only be transformed into mechanical energy through conversion by animals and even then it had to be present in certain forms such as grain and hay. Motion could not be generated from wood. In principle, this only became possible from the early 18th century with the invention of the steam engine.

In the context of the agrarian solar energy system, a transformation of different forms of energy was only possible through bioconversion or altering the use of space. If there was too much wood (= potential heat) and not enough mechanical energy available, the only form of substitution was to fell an area of woodland and use it as pasture. Apart from the use of wind and water power, which played only a subordinate role in the overall energy budget of agrarian society, the problem of energy utilisation was one of alternative land uses.

Here the first and most important characteristic of the agrarian solar energy system becomes apparent: dependence on territory. The overall area of a landscape determined the theoretical total quantity of energy available in this landscape. The upper limit could only be surpassed if it was possible to increase the area, for example by spatial expansion, or through the import of energy fixed in biomass, that is by equivalents of space. There were limits to these imports because of the transportation costs mentioned above. Within a certain territory, specific areas had to be dedicated to each form of energy, which involved a zero sum game: the increase of one kind of energy was only achieved at the expense of another, since it involved alternative uses of a given total area.

From this general dependency on space a series of other features followed that characterised the solar energy system. First, its decentralised nature must be noted. Incoming solar energy has a very low energy density, as a consequence of which energy carriers never occur naturally in a concentrated form. Plant biomass, on which the agrarian solar energy system relies for the most part, principally accumulates dispersed over wide areas. To be utilised by humans it must first be concentrated spatially. This concentration of energy, in foodstuffs, animal fodder or wood, was always associated with a significant expenditure in energy transportation. Naturally no negative balance must arise from this and it was ruled out that transport should consume more energy than the load contained.

These terms limited the overall territory from which energy could be taken for a specific economic location. If it required overland transport, the area would be small. If there were possibilities for employing water and wind power – rafting and sailing – the catchment area could be enlarged so that larger agglomerations could form. Fundamentally, there was no advantage in the size or concentration of settlements or commercial centres, but rather a tendency towards a decentralised distribution of population and commercial facilities across the entire agrarian space. The overall structure was characterised by neighbouring islands of scarcity with the averages of overarching spaces being virtually irrelevant: thus, there could be great scarcity of energy in one settlement even though open spaces still existed in the distance, if the latter could not be exploited with a justifiable expenditure, that is with a positive energy harvest factor. In case of doubt, new small settlements would form again and again, or energy-intensive trades like glass works would move into areas in which sufficient lands (especially woods) were still available.

The same applied to wind and water power. Neither occurred naturally in great abundance, so that extensive facilities needed to be created for their use that quickly reached the upper limit of what could be concentrated in one place. The output limit of a water mill was determined not only by the amount of water available, but also by a technical barrier resulting from the plodding mechanism of wooden devices. In general, it was simpler to build several mills in a sequence with sufficient water power, than to attempt to build one single, very productive mill. The creation of energy-intensive centres of commerce was therefore precluded.

The inherent tendency of agricultural societies to strive towards a stationary state resulted from these fundamental characteristics of dependency on territory and decentralisation. However, the amount of energy theoretically available in an area was utilised with a technical efficiency well below what was physically possible. Here innovations were fundamentally possible in the sense that one could endeavour to improve the degree of efficiency and approach the upper limit determined by physics. This would happen, for example, when switching from an extensive to an intensive method of cultivation. However this was a tough and difficult process. Explosive and long-term processes of 'growth' were only possible as an exception, usually not so much due to technological breakthroughs as to the opening of new land, for instance during the settlement of America by Europeans, which pushed the limit of growth up a few notches. But pioneer phases usually expired quickly in a new stationary state that was again a state of general scarcity of land and energy.

The necessity of managing the energy system, especially by optimising the use of land, rested on this tendency towards a stationary state. This pertained fundamentally to the organisation of the flow, in which the quantity of the flow as such was a given. The core of the problem lay in the appropriate distribution of the available stock, which could not be increased. This distributional pressure constituted a paradigmatic zero sum concept with which agrarian societies ordered their world: wealth required poverty as a precondition, power power-lessness and fortune misfortune. Within this structure a permanent tug-of-war took place over the sharing of a pie, which essentially could not grow. The principle of sustainability was an expression of this central motif in agrarian society: it was the epitome of a well-regulated, disruption-free and permanent balance between a given amount of resources and their stable utilisation. And yet, more and more forces appeared that pulled at the natural bonds that had been imposed upon the agricultural system.

5. The Dynamics of Agrarian Societies

In contrast to hunter-gatherer societies, agricultural societies had considerable leeway for innovation. The carrying capacity of land depended to a much greater degree than in primitive societies on the method of cultivation. The efficiency of cultivation could be improved through a series of measures, some of which have already been mentioned: crop breeding, reconfiguration of growth condi-tions, eliminating competitors, fertilisation, improvement of energy efficiency in transportation and other forms of utilising mechanical energy. The extent of the solar energy flow is on the whole given, but the efficiency of its use can be improved.

Basically, in agrarian society there are always methods of innovation even if these remain within certain limits. Let us assume, for example, that a certain area of wood persistently yielded 100 m^3 wood annually. If the productivity of the wood harvest was doubled, there were two conceivable consequences: either the 100 m^3 wood was harvested in half the time without increasing the overall quantity of wood or it was possible to fell 200 m^3 wood. Obviously, that could only be maintained for a short time. The latter, non-sustainable strategy was quickly penalised by exhaustion of resources. The former, however, opened time for alternative activities to which one might turn in the time gained.

The origin of the transition to agriculture may rather be sought in scarcity, a need to feed a population too large in relation to the available food supply. A successful transition to agriculture could in turn have led to a surplus. This surplus was the formal requirement for social stratification. A differential division of labour that freed entire groups of people from direct work in acquiring their food to be able to exercise an activity considered important enough by those occupied in primary production to make them provide them with food – for whatever reasons, voluntarily or by force – was only possible if more food was produced than the producers consumed. In theory this could happen in hunter-gatherer societies as well, since they had freedom to gather a surplus. It occurred sporadically, for example when providing for shamans. Only agriculturalists institutionalised this division of labour. Only they had a need for truly specialised activities, since they were dependent to a significantly greater degree on the use of tools.

Another important factor also entered the equation. Agriculture must be planned and organised over an annual cycle, or in multi-field systems over an even longer period of time. Farmers needed to keep seed grain, store supplies, maintain work buildings and equipment and care for the state of the soil. Therefore, it was reasonable that significantly more objects, including the land itself, were left at the disposal of the economic unit concerned than was the case among hunters-gatherers. Property rights and the quarrels associated with them were found in every agricultural society and, therefore, social institutions arose that regulated such conflicts. A differentiated legal and state system was encouraged in this manner, and with it the rise of a particular class that held these duties in its grasp. This class had to be fed with the surplus of primary producers.

The creation of a ruling class that lived on agricultural surplus had both a functional and a parasitical aspect. Let us assume a typical situation in which a relatively unstructured peasant society lived in spatial proximity to a hunter-gatherer society. It would be sensible for the hunters (or pastoral nomads) in a situation of scarcity to attack the peasants and steal their resources. Raiding another hunter tribe would be pointless. It was easier and associated with fewer dangers to hunt game oneself than to fight over it with another hunter. War among hunter societies did not occur over prey but for other reasons, perhaps the defence of a territory, but especially for socio-cultural reasons (Keeley 1996). By contrast something really could be gained from farmers. It may have actually seemed less costly and dangerous to attack a village than to hunt wild animals. Therefore, peasants needed to defend their property, so they paid off their enemies with tribute – tribute was first a substitute for plunder, then became institutionalised plundering – or they employed military specialists who were also provided for and who tended to usurp power. Whether tribute, tax or rent:

it followed from the fact that peasants possessed property that was worthwhile plundering that a class could arise which monopolised agrarian surplus.

In principle, there was no difference if the ruling class was recruited from conquerors or if it descended from within a particular society. They protected peasants against robbers by legitimising and sustaining robbery and assumed functions that might be useful to the community. Therefore, peasant societies tended to form social hierarchies and turn into agricultural civilisations. This apparently happened independently on several occasions in history, in Eurasia and in America, which demonstrates that the transition to more complex agricultural societies was contained in the structural principle of agriculture itself (Crone 1989; Sanderson 1995).

In agricultural civilisations not only the fruits of soil and labour could become prey but the labour force itself. If the conditions for the development of surplus were present, an appropriation strategy might consist not only of acquiring the surplus already generated but of increasing it through forced labour. Forms of serfdom that were supposed to limit the mobility of peasants were widespread. Slavery was a common feature of agricultural societies and in historical terms only waned with them. However, a slave was no more likely to produce a surplus than any other worker, since he not only gave rise to acquisition costs but also incurred continuous costs.

Slavery was therefore not profitable under all circumstances. It had a number of disadvantages: labour productivity was rather low because slaves, who were provided for in any case, would prefer to spare their strength. There was a considerable investment in the supervision of their labour. Operation was quite inflexible because in a poor economic situation they could not be sold easily. However, this was juxtaposed by savings in the cost of raising them if adult slaves were bought. If there was monotonous, simple, spatially concentrated work that was easy to supervise, if there was a shortage of free labour and a good supply of cheap slaves, such as prisoners of war, in the markets, then slavery can be anticipated in an agricultural society. However, it could never become dominant because it required that slaves were captured and kept under control. This was the case in many commercially and militarily expanding societies, but slavery always subsided if the supply of slaves dried up.

Agrarian civilisations, of which the origins extend back 5,000 years, exhibited certain similarities. This was even true of civilisations that could not have been in contact. Similar traits to those in Eurasian civilisations are found among the Incas, Mayas and Aztecs, so they can be taken as indications of comparable solutions to related problems. Peasant society, which comprised at least 80% of the population, was always the basis of agrarian civilisations. In addition to this rural foundation, they also produced an urban and commercial sector that was usually geared towards the needs of the ruling class. The governmental apparatus, maintained by the surpluses of the peasant economy, was usually subdivided into three sectors: military, urban administration and public works sector, and the sphere of ritual and religion.

This governmental apparatus was normally concentrated in the towns that could be found in all agrarian civilisations. The typical preindustrial town fulfilled a series of functions closely linked to the system of governance. It almost always possessed a military fortification, an *oppidum*, which was capable of defence. For this purpose special buildings were created, especially the town wall, which could only rarely be omitted, as when a truly effective peace existed in the early Roman imperial period. Furthermore, the town was usually a religious cult centre in which important sanctuaries and priests who served it were located. As a centre of administration the residential town offered employment for a multitude of persons who were concerned with provisioning, tribute collection, armament, jurisprudence and archiving. Aristocrats who sought proximity to the court for the sake of their careers lived in town. Resplendent buildings, theatres, arenas and technically sophisticated monuments that were marvelled at by the peasant population were constructed as objects of internal and external representation.

Foods and raw materials of all sorts flowed into the metropolis to be processed and consumed there. A number of service professions, wage workers, artists, doctors and traders, who formed the mass of the urban population, settled there to provision the governmental apparatus. Finally, the town was an attractive place, a magnet that drew people in from the country who hoped to make their fortune there. Therefore there was always an urban underclass, a half-criminal mob that was feared by the rulers and kept at bay with difficulty.

Another type of town in agrarian society originated not as an administrative centre of the rulers, but as a settlement of armed traders and craftsmen. The trading town located at natural ports, oases and the crossroads of traffic required military fortification to protect trade goods against robbers, or to secure booty from a successful raid. Here, too, palaces were created in which successful trade magnates lived; here sanctuaries and sacred buildings were found, and here larger populations would mass together hoping for employment or subsistence.

The town belonged inseparably to agrarian civilisation. As a trade and occupational centre it appeared to transcend the agrarian principle, but in the end it remained inseparably tied to the agrarian basis. Town dwellers could rarely make up more than 20% of the total population of a given area. From the rural perspective, the town is a parasitic appendage where surpluses are consumed as luxuries which have been extracted from the rural population as rent.

The urban power centre of agrarian civilisation remained dependent on resources flowing in from the countryside and on the loyalty of the rural population. For these reasons it could not do without a broad support base, resulting in specific difficulties related to the principally decentralised character of the agrarian solar energy system: the larger the distance between centre and periphery, the more difficult the task of securing loyalty. Since it was not possible to send out punitive expeditions all the time, a certain legitimacy of central authority needed to exist at the periphery and to be secured on a permanent basis. Consolidation of rule was only possible through concentration of power, but in the end the exercise of authority was tied to a decentralisation of decision-making competence. In a sense a permanent symbolic flow of legitimacy from the centre to the provinces must be organised, mirrored by a material return flow of resources in the form of taxes or tribute. From the perspective of local bearers of authority at the periphery, this return flow of resources to the centre could be seen as superfluous. They tended towards a centrifugal rebellion, towards secession that would always become acute when the centre showed signs of weakness and decay of authority.

Agrarian empires have justly been compared to giants on clay feet that all collapsed sooner or later (Eisenstadt 1967, Tainter 1988, Yoffee and Cowgill 1988). However powerful their concentrated military might may have appeared, however impressive their towns with their temples, palaces and defensive structures were, they yet proved to be feeble when even relatively minor tremors occurred. It is always surprising to learn how small conquering powers were: the army of the Mogul, which brought a large part of the Indian subcontinent into its power in the 16th century, comprised barely more than 5,000 soldiers. A handful of Spaniards caused the Aztec Empire to collapse. Even a private trade company like the British East India Company was able to rule great parts of India. Mounted nomads from Central Asia periodically overran the huge Chinese Empire. 'Barbarian conquest' was a permanent threat to agrarian civilisations. This is an indicator of how poorly these political systems were integrated, and of what a decisive role their permanent shortage of resources and systemic lack of means of communication and transport played.

The collapse that spared none of these agrarian civilisations did not usually mean that agrarian production as such collapsed. Urban centres of authority disappeared, their buildings crumbled, their systems of interpretation were forgotten, their symbolic worlds became incomprehensible. However, below the destroyed civilised apparatus peasant society remained intact. Collapse usually came as contraction: the system retreated to its indestructible core, a small, decentralised rural world, where landlords and shamans took the place of kings and priests.

In agrarian societies, especially in agrarian civilisations, a series of processes would always be initiated that had an essentially dynamic character and strove towards increased access to resources. This must not be understood

as an inherent tendency towards 'development' or 'progress.' Agrarian civilisations could, as far as their technical and economic profile was concerned, remain
stable in the medium term and, in any case, did not tend towards overreaching
themselves. At the same time it cannot be overlooked that certain technical skills
were increased over longer periods of time within their framework, so that the
impression may arise that forces effective in the long term and which could be
understood as 'modernisation' were operative.

No historical teleology need be surmised to understand these tendencies
towards dynamism. Much is explained because agrarian production repeatedly
gave rise to problems, of which solutions had to be attempted. In the first place
this applies to agriculture itself, continually threatened by soil destruction and
struggling with erosion, new pests and nutrient scarcity. It reacted by developing
new methods of cultivation – breeding, fertilisation, crop rotation, draught
power, tools. The same applied to urban trade. For instance, continuous
improvements in metallurgy may be observed in agrarian civilisations. This does
not need to be seen as an autonomous tendency towards progress, but can be
understood as a cascade of reactions to looming scarcity (Wilkinson 1973).

Commercial metallurgical processes relied on the use of exhaustible raw
material reserves. In the beginning those materials most readily accessible were
used, for example copper, which occurs naturally in a pure state. Metal objects
corrode and are worn during use so that they finally disappear, being dispersed
into the environment. Therefore, commercial use of materials was related to a
kind of material entropy: metals were found in ore deposits of a certain
concentration, were further concentrated in the smelting process but were then
diffused during use in a way that did not permit recycling. This meant that in
the course of time new deposits had to be developed, and it could be anticipated
that they would be more and more difficult to access or that the metal content
of the ore would decrease progressively. If a certain level of consumption was to
be maintained in the long term, technical innovations had to occur to make this
permanently possible.

Another dynamic element of agrarian society was its tendency towards
population growth. The cultural population control that was widespread among
hunter-gatherer societies and probably in simple peasant societies due to their
high transparency failed in agrarian civilisations. There were several reasons for
this. One of these was the fact that highly developed agrarian societies reached
a critical population density that favoured the occurrence of serious infectious
diseases, of epidemics and pandemics related to the ecological relationship
between parasitic micro-organisms and their host (see McNeill 1976, 1980;
Cohen 1989; Ewald 1994).

Another factor favouring irregular fluctuations of population numbers
was the mortality gradient between town and country. Preindustrial towns were

usually not capable of maintaining themselves demographically, since the unhygienic living conditions and short transmission routes for pathogens caused mortality to lie permanently above the birth rate. For society as a whole this meant that towns were a kind of population sink: the rural surplus population moved into towns where it eventually disappeared. This meant that farming households in the countryside could tolerate low population growth since there was a chance of outmigration by the surplus.

The result of these processes was that population figures in densely settled agrarian civilisations did not remain stable but fluctuated widely. Wars, famines and epidemics were the permanent, unvanquished curses of agrarian civilisations. Increased mortality was compensated by an equivalent increase of births, but the up and down fluctuations were far larger than was the case in primitive or industrial societies. This increased throughput made cultural regulation of population movements more difficult: if the population was regulated in Malthusian fashion by an incalculable mortality, the control of birth rates by behavioural mechanisms in any case lost meaning. The consequence of this was that complex agricultural societies tended towards permitting population growth and even valued it. Mature, highly developed agrarian civilisations were usually overpopulated, which generally exerted a permanent pressure that could lead to both conflicts and innovations in the resource base.

The agricultural mode of production relied on a permanent inflow of solar energy and therefore it was in principle possible for it to achieve an ecologically steady state. However, this state was far more fragile than that of hunter-gatherer societies. Since agriculture intervened to a far greater extent into natural systems and transformed them, the possibility always existed that it would destroy its own natural basis. The agrarian solar energy system was energetically sustainable since it tapped into a given flow of energy that it simply re-routed without changing its volume or being able to exhaust it. But this only applied to the energy aspect. In terms of materials, solar energy systems were not necessarily sustainable. Here irreversible processes did take place and stock was indeed consumed, soils for example through erosion or salination, or species that were wiped out or deprived of their subsistence basis.

Abundant examples exist of agricultural activities destroying the soil. Agricultural societies were not the ecological ideal they are often presented as being. However, those were often avoidable errors. Many of the sins of ancient agricultural societies that Hughes (1975; 1996) lists could have been avoided if there had been awareness of them or if means had existed to transform this knowledge into action. In antiquity the links between deforestation, goat pasture, erosion and flooding were in fact known (Glacken 1967) and there are numerous indicators in the Roman agronomists of techniques of soil treatment

(Winiwarter 1999). However, it is questionable how broadly this knowledge was diffused and what alternative actions were really available to farmers.

It is permissible to ask if agricultural production as such, that is, independent of its social form of organisation, could remain stable over longer periods of time. Hunter-gatherer societies could, as far as their social and metabolic base was concerned, persist as a physical form of existence as long as their ecosystem remained intact. Its destruction was generally outside their scope of action even if one may speculate to what extent they were capable of an extensive extermination of prey species in pioneer situations (Martin and Wright 1967; Grayson 1980; Donovan 1989; Klein 1992). The capacity to destroy is far greater among farmers, as can be seen in the fate of several cultures of the ancient Near East and pre-Columbian America (Tainter 1988; Yoffee and Cowgill 1988). In the exhaustion of resources and soil destruction the energy system only played an indirect role, insofar as it fed an extension of the range of options. Humans could not destroy the energy system as such because the sun as an energy source was inexhaustible. This cannot be said of other physical parameters. Agriculture linked to metallurgy was probably not possible over as long a period of time as hunting and gathering.

Agrarian society was caught in a sort of dynamic trap: an upper limit was set by its energy basis but it must permanently attempt to reach these limits and even to transgress them, whether as an answer to scarcity or as autonomous cultural innovation. Therefore, it was straining against the natural bonds that the solar energy system had put on it, and as long as there was some leeway it was successful: the limits to growth then merely turned out to be growing limits. As long as it remained within the boundaries of its energy regime, agrarian society faced a situation in which dynamic processes finally ran out or broke down, which might well be associated with catastrophes. The only and epochal escape from this dilemma was to finally break the bonds of this system and to shift to an energy regime of an altogether different nature.

6. Crisis and Transformation

Agrarian societies possessed a complex interdependent structure in which technical/economic, social/political and cultural/ideological systems may be differentiated (Gellner 1988). Each of these constituent systems can to a certain degree be represented separately, while the others may be considered its environment or a part of the framework. It may be assumed that each of these systems could unfold a certain autonomous dynamic which as a rule was buffered. In reality the individual systems were linked to such an extent that they can only be separated analytically. An important achievement of these societies was to enable a stable situation in the face of enormous potential for accelerated cultural evolution. Their high potential for development was artificially reduced, if not halted, because too rapid an evolution always bore the risk that society adapted too quickly to contingent and transient environmental conditions. The formation of cultural tradition can be understood as such a brake on innovation, with the function of steadying or reducing risks of running into an evolutionary dead end.

Therefore, the formation of cultural conventions and traditions implied that a whole range of possible behaviours was either withdrawn from the conscious disposition of individuals or became formalised. The motivation of individuals was shaped in such a way that clear signals were used and interpreted to make behaviour calculable and predictable. Cultural norms and technical and economic knowledge were obligatory so that social cohesion and control over the environment were simplified and standardised.

The stabilisation of traditional behaviour permitted societies to subject a proven profile of behaviour to standardisation. The associated prohibitions to thought and brakes on innovation had the objective of suppressing risky experiments. This in turn involved risks, because a certain potential for innovation must be preserved to allow reaction to changes in the social and natural environment. Therefore, a difficult balance between traditionalism and willingness to innovate with the complementary dangers of rigidification and self-destruction was at stake. The test of these strategies in turn was set in an evolutionary process.

As we have seen, instabilities could arise at different points in agrarian societies. In summary, the most important factors are:

- Production of a surplus, therefore conflicts over distribution and power struggles within and among societies, consequently a 'dynamic history'.

- Necessity of reproducing an artificially created and controlled environment that places a premium on technological innovation.

- Greater growth potential of population; less reliable social indicators for fertility control; therefore larger demographic fluctuations.

- Establishment of information centres as a result of trade and urban concentration, which can lead to a relativisation of cultural norms, processes of enlightenment and, therefore, to accelerated variations within the cultural/ideological system up to the collapse of cultural traditions and the origin of a decided orientation towards innovation.

Many agrarian societies succeeded in institutionally taming the instabilities that resulted from these processes, but success was not always guaranteed. When several such processes converged in a historical situation, the result could be what has been described above as a 'bifurcation'. It must be assumed that the explosive transformation that was initiated by the Industrial Revolution occurred when several such factors culminated in a unique historical situation. The single factors as such, that is the 'causes' that are cited for this development, may be found in isolation in several different historical contexts without establishing a fundamentally new trajectory of development, such as was the case in the Industrial Revolution.

Looking back, the impression that a certain development was directly leading towards a certain goal, that of the present, may easily arise. This may almost be presented as a historical law by theoreticians of modernisation. Thus, we discover a whole series of factors that appear as groundbreakers and preconditions in the search for the causes of the Industrial Revolution. It is not difficult to draw a straight line from the philosophy of nature in the late Renaissance through classical mechanical physics to the machinery of the 19th century. The Industrial Revolution appears to have been co-founded by that revolution in science. Without Newton and Galileo no steam engine, no railway system, no industrial production. On the other hand, isn't it conceivable that even with the existence of Galileo and Newton there could have been an entirely different technical-economic system? Apparently mechanics as such, of which the principles were known in part in antiquity or in China, does not directly lead to the industrial system. Modern science is only one element of a process during which some developments converged that had no effect in other circumstances.

An influential current in the research literature sees industrialisation simply as the technical and economic aspect of a comprehensive process of 'modernisation' that had set in far earlier and almost inevitably had to result in an economic transformation of the old society (see Sanderson 1995). Cultural and mental processes are primarily seen as the core of this modernisation, especially the unfolding of instrumental rationalism and the emancipation of self-interested individuals. Humans begin to calculate rationally, to detach themselves from communities, traditions and superstitions, to take conscious risks and to pursue primarily their own interests, these being understood as economic benefits. From this primary development, secondary benefits follow, like the displacement of descent by achievement, the establishment of universal criteria, such as human rights, and the widespread employment of functionally specific methods.

This process of rationalisation and individualisation had to lead to greater control over nature, better methods of production, rational accounting, capitalistic calculation of utility and an acceleration of technical progress due to new scientific methods. Thus, the phenomenon of the Industrial Revolution is traced back to earlier roots which would be bound to bear fruit sooner or later.

This pattern of explanation has a tradition that extends far back into the 18th and 19th centuries. The historical coincidence of Renaissance, Reformation and European expansion has been understood as the initial ignition of the modern world for some time. Philosophers like Smith, Kant and Hegel, who knew nothing and could not know anything of industrialisation, already spoke of such an origin for the modern world. Influential historical theoreticians like Karl Marx and Max Weber took over this explanation and attached it seamlessly to the origins of industrialisation and capitalism. Britain was always the model: in the period in question a remarkable unity of protestantism, nation state, commercial orientation, technical pragmatism and experimental science took shape, which in retrospect seemed of necessity to lead to the Industrial Revolution.

However, the Netherlands can be used as a counter example. In the period between 1550 and 1750 they were undoubtedly more 'modern' than Britain, if the conventional criteria are employed (de Vries and van der Woude 1997). Of course, they were protestant, they built a nation state with rational institutions and civil self-government, and their economic specialisation was further advanced, especially in transport, which was vital for mercantilism. No country was so massively urbanised. Around 1675 38% of the Dutch lived in towns with more than 5,000 inhabitants, but only 14% in England and 10% in France. The economic system was largely free of state intervention and regulation. Wages were high and capitalist wage labour was the rule. A rational, sober economic mentality was widespread.

And yet no Industrial Revolution originated in the Netherlands. Rather, for Adam Smith the Netherlands were an example of an economy that had already reached its peak, 'which had acquired that full complement of riches which the nature of its soil and climate, and its situation with respect to other countries allowed it to acquire; which could, therefore, advance no further, and which was not going backwards' (Smith 1776, 111). Holland was a mature, i.e. a stationary capitalist mercantile economy. No technical and economic dynamic unfolded, it had exhausted the potential that was available to an advanced agrarian society with a highly developed mercantile and commercial sector. The transition to a new kind of system dynamic in the sense of the Industrial Revolution was not initiated. This leap remained a unique phenomenon that took place in Britain alone.

Wrigley (1988), following Zeeuw (1978), has pointed out that in energy terms Dutch prosperity in the golden age relied on the use of peat. The Netherlands moved at the margin of the solar energy system. If wood is used as fuel, one is forced after a few years to manage the woods on sustainable principles, to harvest the annual yield. With that any 'economic growth', as far as it depends on energy use, automatically grinds to a halt. If peat is used, the supply of fuel is greater than in a woodland but, of course, smaller than with the use of coal. However, sustainable exploitation of peat is not possible because, like coal, it involves the use of a stock whose reaccumulation cannot be synchronised with the speed of its consumption.

Peat is plant biomass that has not oxidised due to high soil moisture, and is conserved by being sealed off from air. Boggy areas where peat forms can be very large. In Denmark they constituted one fifth to one third of the land area. Digging peat is only worthwhile if the layer is more than 30 cm thick and ashes constitute no more than 30% of the dry mass; in other words, if the proportion of earth is not very high. Digging is difficult and unhealthy because of the high humidity. Freshly dug peat must be air dried before it can be burned. There had to be a considerable shortage of fuel before farmers would use peat (Kjaergaard 1994, 112f.).

However, if a country turns to a systematic utilisation of peat, this can feed a brief period of prosperity in energy terms, the end of which quickly draws nigh. According to Zeeuw's figures, the original Dutch peat stock was 6.2×10^9 m^3, of which about 15.5×10^6 m^3 were dug annually during the 17th century. If one considers that peat was already in use before 1600 and that some remained inaccessible, it can be assumed that c. 0.3–0.5% of the total stock was used annually (Zeeuw 1978, 9; 14). These reserves would only have lasted 200 years and that at a level of use that was quite low. Let us briefly calculate the potential. The fuel value of peat lies on average around 15,000 kJ/kg. If 10 m^3 dried peat weighs a ton it is the equivalent of a fuel value of 1.5×10^6 kJ. Therefore, the fuel value of annual Dutch peat production was about 2.3×10^{10} MJ, which is roughly the equivalent of a million tons of mineral coal, i. e. a third of annual British coal production in the 17th century.

That is not much. At the outset of industrialisation British coal production was already around 10 million tons a year, and it subsequently rose by an exponential rate. Such growth would not have been possible in the Netherlands because any massive increase of peat digging would have quickly led to a complete exhaustion of the stock. Therefore, the Netherlands were in an energy dead end, while England was setting out on the path of innovation. It too relied in principle on a finite resource base, but one so much larger than the Dutch that it could be considered infinite. The Dutch example is therefore a counterfactual one for the significance of mental, social and natural factors. Certainly there would not have been any industrialisation without capitalistic orientation towards utility and a 'modern' mentality, but our example makes it plausible that these cultural factors would not suffice without the natural foundations.

Even so, this indication of ecological and natural conditions does not explain how the system-transcending dynamic that led to the industrial transformation of the 19th century came into being. Probably of great significance was a restructuring of the way and manner in which the social synthesis was constructed, with normative integration being supplanted by market integration (Polanyi 1944). The market is probably as old as agrarian civilisation, but over the millennia it was repeatedly overlaid by cultural and political forces and held in check. In modern Europe it managed for reasons that are difficult to reconstruct to emancipate itself from those bonds and develop a dynamic of its own which in turn drew in everything capable of dynamicisation.

The individuals who were integrated into the market's division of labour were guided by an overarching and omnipotent structure of order that could no longer be personally or normatively bound, or only mistakenly. The dependence of individuals on the market and, therefore, on circumstances and fluctuations, on remote conjunctures that could hardly be influenced, manifested itself in incomprehensible price movements. The objective mechanism of economic laws became a second social nature, and the individual's position within the economic structure determined his decisive social characteristics. Social roles increasingly took on the shape of economic functions; the economy eventually developing into the dominant structural level of society. The fate of all other institutional and cultural areas of society depended on their relationship to the economy. This movement of increased market integration had an increasingly subsuming force. It enthralled ever more spheres of life, devoured all cycles of the subsistence economy and cultural norms and unfolded according to an internal dynamic tendency: economic growth became the system imperative.

Instabilities and ecological collapse occurred repeatedly in agrarian civilisations but in the medium term stationary and permanently 'sustainable' states arose. On the basis of the fossil energy regime, a new energy form arose that was founded upon permanent economic growth and therefore was necessarily oriented towards permanently accelerated social, cultural and technical transformation. As far as it is balanced, this balance rests on a dynamic process: the flow of large and usually increasing amounts of energy and raw materials in a manner that was beyond the agrarian system.

Even in agrarian societies a threat of ecological crises and even collapse existed since there was a characteristic delay between the intervention in natural systems and the emergence of undesirable side effects. The inertia of natural processes put the final test of environmental compatibility of certain production processes so far into the future that the learning curve could be catastrophic if not lethal. Several agrarian societies had this experience. In the industrial system there is a far greater discrepancy between innovation and awareness of its consequences. Far-reaching actions are initiated without even the remotest knowledge of what the aggregate long-term result may be. From the perspective of world history, the industrial transformation is a tremendously risky enterprise. If it ends in a dead end, there is a danger, because of its tendency towards globalisation and universalisation, that for the first time in history not only a single society but in principle all of humanity will be affected.

The process of genesis of the industrial system itself is extraordinarily complex and to date barely understood. Ever since observers in the 19th century became aware that they faced a novel phenomenon, many attempts to explain it have been presented. Today many contemporaries, even in the fields of history and economics, consider industrialisation as so natural that they see no need for explanation and are satisfied with empty formulae like 'modernisation' or 'economic development'. However, there are also remarkable approaches that take the problem seriously. (Baechler, Hall and Mann 1988; Hall 1985; Jones 1987; Landes 1998; Wrigley 1988).

In the first approach a series of elements can be cited that are inherent in the agrarian system, of which the specific convergence in Europe led to the birth of a new era in industrial transformation. These may be of varying importance and they may be weighted differently in each agrarian civilisation. They can be listed as follows:

- Creation of social cohesion through a web of relationships, at the centre of which lies the market and the production of goods.

- Institutional preconditions, in the sense that the agencies of political power are understood as service organs of society.

- Universality of private property, i.e. even soil and labour are for sale and traded on the market.

- Emancipation of (scientific) thought from institutional embeddedness, i.e. rupture of the stabilising power of tradition.

- Emancipation of nature-altering actions from social and political restrictions which enables increasing pragmatic subjugation and mastery of nature.

- Emancipation of economic action from cultural restrictions, i.e. establishment of the principle of profit maximisation as the driving motive of individuals and simultaneously as the imperative of the system.

- Rapid population growth, which requires an extension of the basis of subsistence.

- Political fragmentation, i.e. absence of a political power that would be capable of massively disrupting the autonomous economic processes.

- Cultural diversity and plurality in Europe, i.e. great and growing room for politically and institutionally independent communication.

- As a limiting condition: occurrence of easily accessible and easily transported fossil energy.

It is apparent that some of these factors were present in other historical and geographic situations, too, which leads to all sorts of speculation about an Industrial Revolution of the Middle Ages (Gimpel 1980) or the economic primacy of Asia (Frank 1998). Instead, one suspects that it was a specific, highly improbable and, therefore, unique convergence and combination of numerous factors that provided the impetus to industrial transformation in Europe.

No attempt will be made here to provide a comprehensive explanation, particularly since one must assume that the origin of the transformation process lay in a contingent constellation of individual factors that mutually supported each other and cannot be reduced to the prominence of a single factor. It is the goal of this study merely to draw attention to a set of components that have not been properly appreciated so far: natural and ecological foundations and, in particular, the energy system.

7. The Industrial System and Fossil Energy

European societies of the 18th century were still contained within the general pattern of agrarian civilisation, even though they had exploited the potential of the controlled solar energy system in exemplary fashion. Numerous elements of the technical, cultural, scientific and economic state of Europe in the 18th century had parallels in other agrarian civilisations, especially China and Japan. However, in Europe these phenomena stood in a particular combination, which resulted in such a unique dynamic that – seen retrospectively – they would virtually of necessity transcend the agrarian pattern during industrialisation. In this book it will be demonstrated that the occurrence and accessibility of fossil fuels were a limiting condition in this transformation process. Without coal, European societies of the 18th and 19th centuries would have remained agrarian societies, even if they had utilised the innovation potential fundamentally embodied in agrarian societies to a much greater extent.

This emphasis on ecological and economic conditions should not be understood as naturalistic reductionism. From an ecological and natural perspective, to achieve a specific economic effect two different elements are required:

i. A resource, a material or an energy carrier;

ii. A procedure for its utilisation in the widest sense, that is a technical instrument, a social need, a cultural acceptance, an economic structure or a political and institutional framework.

The difference appears trivial but it is necessary to be aware of its consequences in a historical context. Only the coincidence of resource and procedure creates an effect, and for this it is necessary that both elements are present. The mere existence of a resource, such as coal or oil, is inconsequential if it cannot be used. The existence of a demand or a procedure is also useless if the resource is lacking, for example, if it is exhausted. This particularly applies to agricultural societies in which numerous resources (and energy forms as well) have a qualitative character, i.e. they cannot be transformed into one another. The assumption that is valid in today's economy, that everything can be substituted for everything else, simply does not apply in the agrarian context.

For an explanation of the industrial transformation process a general differentiation between favouring and limiting factors is meaningful. A favouring factor takes an active part in structuring a field of events, and therefore has under all circumstances a modifying influence on overall structure. By contrast, the availability of fossil fuels like coal is a limiting condition within this web of factors, since the new industrial mode of production could not have developed without this component. On the other hand, the mere isolated existence of this factor obviously does not unfold the slightest dynamic in a different historical context. Coal existed before humans. It existed at the time of the Palaeolithic hunter-gatherer cultures and in the subsequent age of peasant societies. In numerous agrarian civilisations fossil fuels can be found and are even used sporadically, as in China, Mesopotamia and the Roman empire, but without triggering a revolution of the energy system. It was only in Europe during the modern period when a unique dynamic was constructed that coal suddenly gained its strategic importance. Only then did access to this resource hold the promise of overcoming the energy bottleneck of the agrarian solar energy system.

In the course of the Industrial Revolution all essential facets of life were affected and revolutionised, and this also applies to the energy system. For the first time in history, human beings switched systematically on a large scale to energy carriers that are not permanently renewed on the same scale. The energy basis and precondition for the Industrial Revolution was the use of fossil fuels. They, too, are the product of photosynthetic binding of solar energy – they were created during a period of the earth's history when more plant biomass was formed than was oxidised and lost in respiration. However, during the period when they are being consumed no fossilisation of plant substances worth mentioning is taking place. In contrast to the regenerating biomass, to wind and water energy, fossil fuels are a one-time-only bank of stored energy. In a historical timescale, these stocks cannot be reproduced. Currently about 10 billion tons of fossil fuels are consumed annually on a global basis. However, that is an amount that was stored in 500,000 years. Therefore, it can be said using an economic metaphor that the industrial system consumes not the energy income of the Earth but its accumulated wealth.

However, this wealth is quite large. There has been speculation for some time about how long the fossil energy age may last. The difficulty of this calculation lies in the fact that 'fossil fuels' are by no means homogeneous. It is not the consumption of a defined stock, as if capital of a given size is consumed – there are also qualitative problems. The carbon reserves of the earth's crust are huge but much of it is only available in low concentrations, contaminated with minerals and chemicals or difficult to access. The historical course can be imagined as follows: first easily accessible and high quality stocks are exploited and then gradually the remainder is used. This transition should be associated with rising 'costs' or increased investment of energy within a given technology. If one assumes that 'technical progress' in the sense of cost-neutral improvements of efficiency will take place, then this tendency is counteracted. However, future technological progress cannot be prognosticated and nobody knows future consumption figures, which will be influenced positively by increases in production and negatively by increases in productivity.

Precisely because of these qualitative aspects, an abrupt collapse of the fossil energy system can be ruled out. It can be assumed that costs will rise in the long term but this will place a premium on processes of adaptation: energy conservation, improved prospecting, and substitution. All this may mean that energy will become scarcer and more expensive so that behaviour that is associated with low energy usage will be rewarded. Overall, it will involve a continuous process of adaptation within which there will be crises but no dramatic end. Ecological problems in the narrow sense will probably result from the emission side rather than the resource side.

At the end of the 20th century, global coal reserves (the known stocks that can be mined with available technology) were estimated at more than 10^{12} t. The current use of coal lies at 4×10^9 tons annually. The reserves will last another 250 years at constant usage. If one looks at coal resources (the estimated total stock), which lie around 10^{13} t, the time frame is extended tenfold. Nobody knows what part of these resources can be mined at reasonable cost and it is also unknown how coal consumption will develop in the future.

Today the annual consumption of oil and gas is about double that of coal, while oil and gas reserves amount to only 5% of coal reserves. For this reason alone it may be expected that coal consumption will rise in the 21st century. Furthermore, it should be considered that the per capita level of energy use in the USA is about eight times the average level of the global population. If the desired 'development' of the Third World should come about, this would surely be linked to rising energy use, even if in the future increased efficiency in the sense of 'energy saving' may be anticipated. The fossil energy age will probably still have a duration of several centuries.

If the use of fossil fuels is represented graphically over long periods of time, the image of a spike on the time axis appears (Thirring 1958, 218; see also Cipolla 1962, 59). Up to the beginning of the Industrial Revolution consumption was virtually zero, a short, steep increase to a maximum occurred, which will then probably be followed by a somewhat flatter ascent and consumption will tend towards zero again over the next millennium. From a very long-term world history perspective the image of a short, at most thousand-year fossil fuel interlude emerges.

From this it becomes obvious at a glance why the Industrial Revolution is associated with a colossal acceleration of material consumption in the sense of 'economic growth'. Suddenly and very quickly much more energy became available than could have been provided by the agrarian solar energy system. Global consumption of fossil fuels has grown by a thousand fold since the beginning of the 19th century, which amounts to an annual growth rate of approximately 3.5%. Industrialisation depends in terms of energy on two closely linked processes: enormous reserves became available and their exploitation grew exponentially.

Humanity suddenly had fuel at its disposal on a scale that far exceeded what was available in regenerating biomass. This is unique in historical terms if not in the entire history of the earth. An agrarian society in a pioneer phase settling previously unused space could enter a similar situation of energy superabundance. Primeval forest in North America, for example, contained biomass formed over a period of 300 years. If this forest is clear cut, settlers can dispose over a yield that is three hundred fold the sustainable amount for this area. However, they are only able to cut down the primeval forest once, after which they are forced to turn to sustainable methods or move on. It is similar with agriculture. Short-term practices are conceivable that permit producing more plant nutrition than the soil will permanently yield, but the actual carrying capacity of the land will eventually come to bear with definite yield reductions.

The use of fossil fuels by industrial society is an expression of this kind of pioneer phase, but with the difference that it may last several centuries. This is rather long in biographical and historical dimensions. However, from a world history perspective it is not. Fundamentally no social structure can be built on the basis of a consumption of fossil fuels that could have a similar duration to agrarian society, or even to the agrarian civilisations that after all reached an age of 5,000 years. Industrial society, which relies on fossil energy, is ultimately only a transitional society: sooner or later a shift to the utilisation of other sources of energy will be necessary. This could be nuclear energy, which can use almost inexhaustible reserves of uranium in advanced fast-breeder technologies, with the possibility of a fusion reactor that would solve all energy problems. The other alternative would be a restriction to the use of solar energy again, which is possible today with a completely different degree of efficiency than under the conditions of the agrarian mode of production. There is no doubt that far more people could be better fed with an industrial solar energy system than was the case with the agrarian solar energy system. But if we consider that about 750 million people lived in the world around 1750, and that there will be more than tenfold that number by the middle of the new century, there is understandable concern, in the face of expected environmental difficulties and threatening raw material scarcity, whether in terms of pure solar energy that number will ever be able to participate in the flow of material and energy to the extent customary today for the inhabitants of the core regions of the industrial system. However, it is not the task of this historical study to deal in detail with these difficult technical and economic problems.

An important characteristic of the fossil energy system was and is energy superabundance. Many characteristics of hunter-gatherer and agrarian societies are attributable to energy being in short supply and the need to be very economical with it. This was already a principle of organic evolution. The enormous abundance of energy in the industrial system led to the formation of behaviour that appears absurd from an energy point of view. While settlement structures and forms of space utilisation in solar energy societies were arranged according to the principle of a minimisation of transport, transport became almost free once the mineral oil economy prevailed. In the expansion of traffic systems, energy consumption plays virtually no further role. Transport infra-structures take a disproportionately large share in the formation of towns and regions with accessibility and speed at centre stage. Any attempt to establish a technically improved solar energy system faces the enormous problem that the result of energy superabundance has literally been poured in concrete in the meantime.

Another example of such an inversion is agriculture. As we have seen, the Neolithic revolution ushered in decisive technical improvements on the Pleistocene solar energy system. Preindustrial agriculture was a system designed to transform solar energy into forms that were useful to humanity. With the introduction of fossil energy the character of agriculture has fundamentally changed. Agriculture is no longer a part of the energy system, but only serves the material metabolism: it transforms carbon dioxide, water and minerals into nutrition for plants and animals, but requires an external source of energy. It stands to reason that this is only possible as long as fossil fuels are available.

The Industrial Revolution is a singular phenomenon in world history because of the nature of its energy base. The traditional agricultural system is not the ecological idyll it sometimes appears to be. Agriculture is fundamentally precarious in ecological terms because of its relatively short-term intervention into highly complex natural systems. Compared to the ten thousand year duration of the agrarian regime, the industrial system appears to some observers as a one time, short fling in which a treasure gathered over many millions of years is being squandered in a few centuries. This applies not only to fossil fuels but also to the concentrated occurrences of minerals that are exploited and diffused with its help, and to ecosystems and species that fall by the wayside. Perhaps a horrible hangover will follow this fling.

The temporally limited character of the fossil energy system may cast a shadow on the expectation of boundless material progress which has been built up in the course of its expansion into large parts of the world. However, this dark perspective is largely irrelevant in an explanation of the transformational dynamic that has held the preindustrial world in its grasp for the last two hundred years. Human beings do not live in an anticipated future but in the real present. Futuristic forecasts that cover long periods of time are therefore seen, by and large, to be of very little relevance. It is only possible to work with reserves if they are comprehensible, that is if they are relatively small. Within the agrarian system it was possible to make do in a 'sustainable' manner with available resources because their volume was transparent and overuse was readily noticeable. The enormous amounts of fossil fuel that suddenly became available burst the bounds of expectation. There is no human intuition for the use of such huge quantities – it comes more easily to consider them infinitely large.

To understand the processes that were and still are made possible by the energy transformation, the future dimension that has at bottom only prognostic but no empirical value may be ignored. Humans have experienced the growth phase that was fed by the new energy system as a permanent, almost natural process, more or less a matter of course . Often this is still so today. The energy transformation carried large areas of economic and mental reality along with it. Explosive change, growth and transformation have become the signature of the age. My purpose here is to understand better the historical origin of this process.

II

Forest and Wood in Preindustrial Germany

1. Natural Foundations

For any stage after the mastery of fire, it is meaningful to distinguish between energy flows that are mediated by the human metabolism and those forms of energy use that take place outside biological metabolism. In hunter-gatherer societies these different functions are not permanently assigned to a specific part of the habitat: hunting, collecting fruit and gathering fuel occur in principle in one and the same area; picking wild fruits is no different from collecting dry wood for the hearth.

By contrast, agrarian societies undertake a functional division of their economic territory. In central Europe it is generally tripartite: it consists of arable, pasture and forest, which are in turn associated with metabolic, mechanical and thermal energy respectively. This functional division becomes necessary when the occurrence of any of these is not enough to permit the simple appropriation of natural goods necessary for subsistence, and human beings must take care not to exhaust the productive capacity of nature. The transition from mere gathering to a reproductive economy is fluid. It does not occur evenly and simultaneously in all areas at once. Types of gathering or a pure 'harvest economy', in which the soil is not cultivated but care must be taken that fruit-bearing plants are not damaged and that natural seeding can take place, occurred for a considerable time next to intensive agriculture, which subjects the entire vegetative cycle from seeding to harvest to human control or 'colonisation'. However, the historical tendency was to draw ever more areas into the reproductive cycle. The forest finally became the last component of the agricultural biotope to be used 'sustainably' and managed actively.

After the last Ice Age, central and western Europe was not forested, but a steppe. In the post-glacial period, which began about 12,000 years ago, the forests advanced into the steppe. Climatically it was probably a little colder than today, so that hardy birch and pine were the pioneer plants. About 8,000 years ago average temperatures were about 2–3 °C higher than today; and 7,000 years ago, when the first Neolithic farmers arrived in central Europe, the land was almost completely covered in forest, with the exception of moorlands, which took up 15% of the area. These 'primeval forests' were not particularly old, nor were they ecologically particularly stable, but subject to larger climatically induced fluctuations, which meant that their species composition changed independently of human intervention. Initially, anthropogenic changes were nothing that could be differentiated in principle from 'natural' developments in quantitative or qualitative terms.

It is impossible to say how the 'natural' vegetation of central Europe would look without anthropogenic changes. We simply cannot say what the 'natural' state at the beginning of the neolithic transition might have been: climatic changes have taken place since then that would have caused shifts in the composition of the vegetation. However, it is undeniable that agriculture led to a profound transformation of the natural ecosystem. Agriculture always implies clearing of forests. Today it is estimated that 2000 years ago the 'primeval forests of Germania' only covered 20% of the land, the arable covered another 10–20% of the land and the remainder consisted of productive forests and woodland pastures (Jäger 1994, 81).

Clearance accelerated soil erosion, which led to the formation of characteristic flood plains. Increased sedimentation is virtually an indicator of agricultural production. Furthermore, it must be assumed that some species disappeared or became extinct due to changes in their habitat even before the modern age. On the other hand, it is likely that the establishment of pasture increased the differentiation of the ecosystem, with hedges, ponds etc., so that more varied ecological niches were created than would have been the case without human beings.

There is no consensus in the literature on whether Neolithic agriculture began in parts of Europe that were not covered by forest, or whether clearances were required from the start (Darby 1956, Smith 1978, Lüning 1988). It tells against the first proposition that it was precisely the unfavourable areas in terms of climate and soil that the forest had not conquered. Therefore, they were not well suited for agriculture either. On the other hand, the effort in clearing the original forest was very high if one considers the technological state of early agriculture. There are many indications that slash and burn with an only temporary use of cleared land (shifting agriculture) initiated the forest-changing activities of human beings (Küster 1995).

This is how slash and burn proceeded: The bark was peeled off trees or pierced in a circle near the roots, which caused them to die off. Small trees were felled with a stone axe. When the wood had dried, the forest was set on fire, the ashes were then hoed into the ground and seeding of agricultural plants followed. In the first year this resulted in quite good yields because the ashes fertilised the soil and burning destroyed weed seeds. After the third to fifth year yields sank so quickly that farmers gave up the cleared land and moved on. The required area was huge and stable, but permanent settlements could not form.

The forest as a component of the agricultural biotope

In the end it was a combination of agriculture and animal husbandry that permitted humans to settle for a longer period, since they then became able to use the soil in a sustainable manner. The typical farming biotope of preindustrial Europe appeared as follows: the cultivated areas served for the cultivation of food and useful plants. In the first place farmers needed to store the inflowing solar energy in a form that could be used by the human metabolism. For this purpose grain was cultivated. Furthermore, crops like hemp, flax and the dye-plant madder were grown. Another portion of agricultural land served as pasture. Oxen, cows and horses were in the first instance work animals but were also kept as suppliers of protein. Sheep provided wool from which higher quality clothes were made. Furthermore, animals produced dung, with the help of which nutrients could be transferred from pastures to the arable.

Finally, the third component of the agricultural economy was the forest. It was the least intensively used, forming a residual and reserve area, but it also served several functions at once. In the early Middle Ages its structure may be envisioned as follows. Animals were pastured among the large trees that stood in immediate proximity to small fields. This area of *Hutewald* resembled a landscape park, with huge expansive beech trees and oaks among which grass and shrubs grew. These trees were not used for firewood, but were fruit bearing. Pigs were fed on acorns and beechnuts, which these trees only produce abundantly at a rather advanced age. This open forest eventually shaded into a more primeval forest that supplied fuel but was also pastured extensively. Until the introduction of the potato in the 18th century, acorn mast was virtually the only fodder for pigs that were driven into the forest. It was therefore obvious early on that oaks needed to be spared while felling wood for fuel, so that their proportion tended to increase in the forests.

Expressed as a material flow, the following happened when pigs foraged in the forest: pasturing pigs consumed biomass that was produced by trees and finally transported it back to farms. These were not only carbohydrates valuable for their energy content but also nitrogen, phosphorus and minerals. These substances were partially transformed into the pigs' own biomass and partially carried to the stall in their intestines, where the farmer could shift them together with other faeces, plant and animal refuse as fertiliser into garden and field. Energy and nutrients were transferred on balance from the forest to the field. Preindustrial agriculture was apparently dependent on this transfer: the nutrient losses that occurred because of accelerated soil erosion after the harvest, the lack of plants with deeper roots and the export of plant biomass to sites of human consumption, from which minerals could only be partially recycled to the location of the plants, would have led in the long term to soil exhaustion. However, the extent of this transfer should not be overestimated, since the pigs remained in the woods most of the time.

Trees require fewer minerals than crops in a given time period per unit area because they are not entirely removed from their location after the growth cycle, but phosphorus, nitrogen and trace elements can be circulated on site. Old, rigid wood releases a large part of the minerals contained in it to younger parts of the tree. Of the annual nitrogen consumption, 80% is bound in the foliage. After leaf-fall in the autumn, microbes break it down and the roots can take up the substances contained in it again. Therefore, the necessary replacement of lost substances is small in the forest as compared to the arable.

A stable and naturally fluctuating ecosystem, in which humans only exist as hunters and gatherers, knows almost no imports and exports of minerals, since these circulate within the relevant biotope. In that sense an almost perfect 'recycling' of minerals takes place. Only a small part of the biomass is removed from the site in which it was formed, through feeding by animals or washing away. Only oxygen, carbon and hydrogen circulate on a large scale through the atmosphere and hydrosphere. The small loss of minerals can generally be replaced in the long term by the decay of rocks.

Things are fundamentally different in an agriculturally exploited ecosystem. If the plants are removed from their location after the harvest, this means that mineral substances stored in them are exported to human consumers. Because of this forced export, agriculture can only be carried on permanently in a given location if the required substances, especially phosphorus and nitrogen, are returned to the soil. This is basically only possible when the ground is manured with animal and human faeces. During this process a retransfer of substances takes place: it is not adding anything to the arable, it merely goes towards replacing what has been taken out. This can never lead to complete recycling because substances are always lost during consumption in the same way as they are washed away by flowing water. To the extent that the decay of rocks or input from the air cannot compensate for these losses, agriculture is always threatened by soil depletion. However, the transfer from forest to field can slow down this process considerably and even stop it entirely.

In particular, the nitrogen budget moves within a restricted frame. Loomis (1978) estimates that 20 kg N/ha arable were routinely removed with an annual grain yield of 1,000 kg/ha in the Middle Ages. Natural processes could make up this deficit. Nitrogen oxides are continuously formed in the air by lightning. In association with dust and animal faeces this could amount to 8–12 kg N/ha. If nitrogen-binding bacteria also introduce an annual amount of about 4–10 kg N/ha, then the budget is balanced even if a nitrogen loss of 10% through wind and water erosion is anticipated. Braudel (1974, 80f) assumes a grain yield of merely 500–600 kg/ha for France in the Middle Ages. The deficit would then only amount to 8 kg N/ha, which can readily be compensated by natural imports. From this perspective, fertilising in the sense of a focused nitrogen transfer from extensively to intensively used areas in the Middle Ages is not required.

However, the estimates of Loomis refer to two-field rotation. The budget changes in three-field rotation, since no transfer to the field through seed takes place in the pasturing phase. Mere recycling is involved with animal faeces, not a net addition. If the attempt is made to increase the yield, farmers are soon caught in a nutrient trap: every increase in the yield is paid with a net export of minerals so that the necessity of manuring arises, effectively the use of fields further away.

Nutrient transfer from the forest could provide a solution. It is possible over a longer timescale because trees require fewer nutrients in a given period and because in consequence of their longevity there is a chance that new minerals will be released. Furthermore, the roots of trees act as nutrient pumps that make it possible to draw trace elements from greater depths. Due to this 'hardiness' of trees they can grow in areas that cannot be used by agriculture because of its low yield.

Nutrient transfer from the forest to the field occurred not only with pig pasturage but also through more direct methods. Humus-rich soils were taken from the forest to fertilise gardens, vineyards and grain fields, while foliage was cut for cattle. Even felling and removing wood from the forest is a form of substance transfer. Chopped wood, which consists of more than carbon-hydrogen compounds that are transferred through the atmospheric cycle and nitrogen that is taken directly from the atmosphere by forest soil micro-organisms and made available to the tree, is removed from its natural location with no care being taken that the substances contained in it are returned. In an intensive economy tree plantations must be fertilised. Where this does not happen – that is throughout preindustrial forest usage – there will be growth stunting in forests if excessive amounts of wood are removed. The nutrient transfer from forest to field or generally to the human consumer must not be too intense, or over the long term it reduces the state of the forest.

In general wood was used for three purposes, which can be differentiated as follows:

- as construction material

- as a raw chemical material

- as a heat provider.

It can be said with some justification that wood was the key resource of preindustrial economy (Gleitsmann 1980). The supply of wood was as important to the agricultural economy as was that of other plant products. The forest must be understood as an integral component of agriculture, rather than as an alternative to it.

This should not obscure the fact that from the perspective of the land-taking farmer the forest was first seen as an unruly 'wilderness', as 'enemy of cultivation', and not as a constituent part. First the arable and pastures had to be created, then the forest could be seen as a complement. A fundamental problem existed here: more arable and more forest products were needed in proportion to population growth, in other words the demand for wood grew at the same rate as forests decreased. If this process is considered more closely, it is essential to view arable, pasture and woodland as alternative forms of land use that were in principle substitutable to some degree without one form displacing the other entirely. It was important to create a sensible balance between them.

We have seen that there is a transition from primeval forest to its clearance and transformation into arable – from outright dense forest through different forms of agricultural usage to open field. A forest in which pigs or goats are permanently pastured or from which humus is taken provides lower wood yields than a closed forest managed for forestry. If the forest is protected against other forms of use for the sake of an increased firewood supply, then it is essentially a shift in the balance between forest and field in favour of the former. There is no essential difference if the forest is closed and instead a field is provided to grow fodder beets or potatoes to fatten pigs, or if the same area is reforested and the pigs are driven in for the acorn mast. The difference is merely of a quantitative nature: pig forage in the oak forest requires a far greater land area than the growth of fodder plants. In both cases agricultural land area is lost that could have been used for the production of human food. A prohibition on pasturing pigs in the 'public forest' without compensation means that the structure of land use has to change and the whole farming operation becomes more difficult.

Forms of forest use

Let us consider forms of forest usage somewhat more closely. First it must be noted that the yield of different species on similar soils varies considerably. There are exceedingly fast-growing species that will never reach a very old age such as willows, poplars, birch and aspen, and very slow-growing trees with high life expectancies such as yew, oak, lime and chestnut.

Wood	60 years		100 years		120 years	
	Height	m³	Height	m³	Height	m³
Pine	15.4	308	21.5	404	23.0	430
Spruce	14.2	428	25.0	739	27.5	806
Fir	12.2	315	23.0	784	26.5	934
Beech	16.9	274	23.0	489	25.0	579
Oak	16.2	244	22.8	413	25.2	482

*Table 4. Heights and wood volume per hectare of different tree species
on average soils*

(Hausrath 1907, 20)

At first sight it looks as if there are great differences in the photosynthetic efficiency and biomass production of different tree species. However, the picture changes when dry matter is considered. It increases annually by about the same amount for all tree species. The different mass yields are differences in specific weight. 'Since the combustion value of wood is proportional to its specific weight, it can be said that all kinds of wood produce the same amount of fuel and, that if we merely wanted to have firewood, the choice of wood type is irrelevant' (Hausrath 1907, 20).

That certain species of trees were preferred by the preindustrial economy cannot be a matter of their combustion value. It can be observed that until the 13th century deciduous forest displaced coniferous forest. Then a period with an increased proportion of coniferous forest began; in the 18th century deciduous woodland was favoured, and since the 19th century coniferous forest has been on the rise again. At least the first part of this process cannot be ascribed to direct intent in forestry. The trees most useful to foraging pigs were oak, beech and chestnut. These trees have heavy seeds that do not favour their distribution in competition with the light, wind-borne seeds of most conifers. But they have an important selective advantage: they can sprout from rootstock and, therefore, are less sensitive to damage and less susceptible to insects and fungi. After slash and burn clearing, deciduous trees recover faster than coniferous trees and this may have unintentionally favoured their wider distribution. From the 14th century, methods of conifer seeding were developed and since then the share of these tree species, especially pine and spruce, has increased again. They then replaced oak trees that were predominantly useful as fruit bearing trees producing acorn mast.

Until the late Middle Ages there was no actual forestry. Wood was taken from forests but fruit-bearing trees were spared as much as possible – a practice that was regulated at an early date with prohibitions. But many more orderly management forms under a natural growth regime have resulted. One can imagine that permanent pasturing, especially of sheep and goats, strongly discouraged young growth. If the routes to fellable wood were not to become intolerably long, young shoots had to be protected even in the midst of a superabundance of forests and woods. It was in the interest of the farmers themselves to establish these 'nurseries'. However, this had consequences: 'The axe strokes of a given year had to fall in one place or forest pasture would have had to be abandoned – otherwise, protection against livestock would have been impractical. Sequencing of annual cuts was also in the interest of pasturing, since a discrete area would become available for it' (Hausrath 1907, 47f.).

The result of this process was the so-called *Mittelwald*, or composite forest: if a section of forest was felled, some trees were left standing, whose seeds could develop a new stand. These seed trees or staddles could be felled later with supernumerous young trees, so that younger, thinner trunks could be used as firewood and older seed trees as timber. At the same time the area did not have to be reforested by hand: natural seeding was relied upon. This management form established itself in central Europe no later than the 15th and 16th centuries and was dominant until the 18th century. There was some difficulty finding the right relationship of staddles and young trees. If there were too many staddles, they shaded the young growth too much, but if there were too few, seeds were lacking. Furthermore, the seed trees that formerly grew among other tall trees were very susceptible to wind damage. A series of forestry techniques was developed to make up for these disadvantages, such as shelter wood felling, leaving a wind break, border seeding etc. These techniques are of no further interest here.

One problem of the composite forest was that isolated seed trees easily developed lateral branches, expanded in breadth and therefore did not develop the long, straight trunks required for construction. For this purpose, the high forests were developed: the stands were arranged and planted so that only trees of the same age and species stood beside each other. These trees could not expand laterally. Thus, branchless, tall and evenly grown trunks were raised for use as ship's masts, beams etc.

For pure fuel supply, coppicing was developed early on, often combined with various types of field management. Depending on the region, this economy consisted of coppicing with field crops or forest/field alternation. Only decidu-ous trees were employed. They were felled in winter, the better pieces taken away as firewood, the poorer ones spread on the forest floor and dried. The tree stumps remained. In the spring the left over, dry wood was burned away and the ashes hoed into the ground, then spring rye was sown and in the next year buckwheat or a similar grain. After two harvests the yield was so low that the land could only be used as pasture, but the cattle had to be prevented from feeding on the fresh shoots that sprouted from the stumps. These shoots suffered considerably from pasturing and frequently such trees died, but if the coppice was protected from cattle and game for one or two years, the forest regenerated. Cultivation was rather advantageous to coppices since it loosened the soils.

This form of coppicing permitted agricultural use of soils that were too infertile for permanent cultivation. Within the short cultivation period the bulk of minerals and nutrients that trees had accumulated over a period of 20–30 years was transported away from the woodland. This was only possible because woods were less demanding than crops. After six years the coppice had regenerated to the point that it could be used as pasture again. After about 20 years the rods had reached cutting age and the cycle began again. This type of operation was employed where a large and stable demand in firewood existed, especially for iron works and salt works. The twenty year old stands, especially beech trees, were best suited to charring, and the charcoal generated from them was of a better quality than from larger and thicker stems.

When coppicing was introduced for a constant consumer of firewood or charcoal, it was advisable to subdivide the entire wood into enough compartments so that year after year an amount of the same size could be cut. How this was done may be illustrated with a simple example. If one assumes a coppiced area of 20 hectares, it would be divided into twenty areas of one hectare each, and every year one of these smaller areas would be cleared. The annual wood growth amounts to about 5 m^3/ha, so that the entire area would yield 100 m^3. This amounts to the stock on the oldest part of the coppice, which had been felled the previous time 20 years before. If all the wood of the oldest area was harvested, exactly the amount that was regrown in the total area in one year was harvested, i.e. the operation was sustainable.

If wood consumption of a particular trade is known, it is possible to determine the area required for a permanent supply of firewood for this trade. It makes little sense to base this on the possible wood stock of the forest, because it could only be cut once. A stationary energy supply means a constant inflow of usable energy. If a certain trade (or a relevant group of small-scale consumers, such as the citizens of a town) maintains a constant consumption over a longer period, we can only take the annual yield of wood, as it results from coppicing, as the source of its fuel supply. If we want to calculate the area needed for a stationary energy supply, we will take the annual increase per hectare as the basis and divide the required quantity of wood by it to obtain the required area. As we have seen, the type of wood plays only a subordinate role since the combustion value of dry wood is proportional to its specific weight and woods that produce a large mass in a given period of time have a low specific weight. Therefore, if we assume an annual increase of 5 m³/ha, the value is based on the average specific weight of wood. Woods produced under the same conditions of soil, precipitation and temperature have approximately the same dry weight, that is, they are little different in combustion value, but rather comparable, although certain subsidiary conditions – humidity, wind, warmth, minerals – may have a differentiating effect.

It should be pointed out that equating different types of trees is only applicable to their use as firewood. There are significant differences in other types of wood usage. This applies with timber, where qualities such as hardness, elasticity, vulnerability to rot and so on are important, and if the wood is for construction purposes, working qualities, weight etc. are also critical. Here, too, the actual volume of the wood is often more important than its type. The chemical properties of wood are also different. Thus, not all kinds of wood can be used for iron smelting with charcoal. The further charcoal is removed from the ideal of pure carbon, i. e. the more other chemical elements it contains, the more difficult iron smelting becomes.

The chemical properties of wood were particularly important in the production of potash. Potash contains a high content of potassium carbonate (K_2CO_3), an important raw material that is particularly important in glass production. To produce potash, wood ashes were boiled, the solution was then strained and finally the lye was steamed. On average a cubic metre of wood yielded a kilogram of potash (Rubner 1967, 26) so that the stationary production of a ton of potash required 200 hectares of forest. However, it must also be considered that different woods have a different potassium content so the kind of trees grown in a territory is relevant to the actual area requirement.

Spruce	0.45
Beech	1.45
Oak	1.53
Willow	2.85
Elm	3.90

Table 5. Potassium content of different wood species

(Parts per thousand. Wilsdorf 1960, 15)

Only on particularly potassium-rich soils would it be possible to operate continuous plantations of trees that store much potassium. Potash was therefore preferably burnt from supplies that were old and for which there was no other use because they were too far from other sites of consumption or transportation routes.

The transportation problem

The transportation problem was critical in the use of wood as energy. Wood only grows spread over a large surface, so that it first has to be collected and concentrated before it is transported to the consumer. For that reason the required expenditure of energy to carry a particular quantity of energy to its destination is very high in transporting wood. On land wood could only be transported over a relatively short distance if the energy yield factor was not to become negative. To determine this balance was not easy in preindustrial society. A clear distinction was drawn in the type of area usage. From the moment that a larger pasture was needed to feed the horses that brought wood from a more distant forest than would have been required for reforesting with a coppice close to a settlement, the energy budget was negative. This would have only had an effect in the long term; in the short term and especially seasonally the balance could be allowed to be negative. Price relationships did not necessarily reflect energy budgets. A horse that had to be maintained anyway because it was needed to cultivate fields in the tilling season could be used for wood transport in winter even if that particular energy budget was not positive.

The mechanical effort that the horse was capable of expending could only be transformed into usable heat energy if it transported fuel. The energy costs of the wood that was burnt at its destination not only contained the cost of the inherent chemical energy but also the amount of energy that the horse used above and beyond its basic metabolic expenditure. This energy had to be returned to the horse in the form of plant biomass (grass, hay, and oats). The area requirement thus included not only the wooded area from which firewood was removed but also the pasture for horses.

High energy expenditure resulted from methods of land transport. For this reason, skids or so-called *Riesen* were constructed on mountainsides. They were constructed as close as possible to the site of the cut and ran downhill to rafting water. Such a slide consisted of six *Ries* trees set into a semi-circular shunt in the Alps. The trunks that thundered down it destroyed it in a relatively short time and ripped it into useless splinters. Attempts were made to reduce wear and tear by pouring water over it, which reduced friction, particularly in winter when it froze, but still slides used up a considerable amount of wood. In this case the area from which the slide trees were taken had to be added to the total firewood area.

The construction of special wood paths (draw paths) only began relatively late, in many areas not until the 18th and 19th centuries. The investment in work and materials was very high. It must be remembered that the paths had to be made with pick and spade. This, too, consumed great quantities of wood. Not only were swampy areas bridged with stick dams, but in the mountains roads had to be supported with wood on steep and rocky slopes since rocks were only rarely blasted because of the high costs of gunpowder. Also, many wooden bridges were required over precipices and rivers. Dedicated wood paths that served no other transportation needs were only constructed when other wood supplies were threatened with exhaustion (Bülow 1962, 116).

In contrast to the land route, rafting wood was associated with little energy expenditure. River water was flowing towards the sea independently of whether it carried wood or not. But even rafting had problematic aspects. Timber stands that were to be transported away on the water were not always near a river that carried sufficient water. To make use of the beneficial water energy, a so-called *Wag* or *Wog* (weir) was set up. The cameralist Johann Heinrich Gottlieb von Justi described how it worked:

> There is not a stream so small in the woods that it could not be used for wood rafts. A dam is made to let the stream swell as much as possible. Then the wood is thrown into the swollen waters and the dam is opened; thus it is carried on with the swollen waters for up to an hour and a half; then another dam must be located where the water will swell again; and thus the rafts are carried on until the stream enters a river where rafts have no difficulties (Justi 1761, 462).

It is readily apparent that this procedure requires an enormous effort in materials and work and affects large areas. Even in the rivers there were problems with rafting wood, particularly if mills also used the water. In that case complicated and costly protective devices had to be set up in the water to ensure that the floating wood did not damage mill wheels. In a way, mills and rafting were competing uses of a watercourse. Special devices also had to be built to land the rafted wood so that the drifting wood could be collected again. These devices were, of course, made of wood, so the losses incurred had to be deducted from the overall wood harvest, which again enlarged the required forest area. Even rafting was associated with unwanted wood losses. About 25% of the wood mass was lost due to friction, splintering etc. (Wilsdorf 1960). Even if no actual rafts were constructed, as was the case with firewood, the work involved was considerable. It was prompted not least by the desire to prevent wood theft:

> Stern laws would prevent theft in a matter that is so favourable to the entire public as a good wood price. Even if such were unavoidable, there would still be

more favourable transport if a man was positioned every quarter hour to prevent blocking of the wood and theft (Justi 1761, 463).

These transportation problems become most acute if wood has to be transported to its place of consumption, that is, if the consumers are for some reason dependent on using the required energy carrier at such locations as smelters, salt works, mines or towns. In other forms of consumption production may be shifted to where there is sufficient wood. This was particularly the case with glass works, where the most important raw material was potash produced from wood. If we assume that a hectare of relatively untouched mixed forest that cannot be used for another purpose carries about 600 m³ wood, then 600 kg of potash can be burned without any worry about sustainability. This amount of potash is much easier to transport than the amount of wood from which it was produced. Burning ashes became a form of use for forests that were otherwise hardly accessible. The same also applies to the production of pitch, resin and tanning bark.

Transport conditions and transport costs were a major problem in the context of the agrarian solar energy regime. Bairoch (1993, 60) estimates preindustrial transport costs in terms of grain: it cost 3.9 kg cereals to transport 1 ton of goods over 1 km by cart, 0.9 kg by river or canal and only 0.3–0.4 kg over sea. This exemplifies the energy limit of overland transport, especially for bulky goods as wood which would be felt in the economic sector as prohibitively high costs. When we assume the price of 1 ton of wood to be about 1% of that of grain, wood prices double every 2–4 km of overland transport, or, to put it in a different way, for each kilometre the price of wood will increase by 40% if transported over land, 10% over water and 3% over sea.

In general, it can be assumed that overland transport of quality wood was not worthwhile if the ride took more than six to seven hours, while the limit was already reached with firewood after three to four hours. This represents a distance of 15–30 km, but only if the state of the roads, especially in winter, when they are frozen, was good. A settlement that was not on a raftable river was only able to draw wood from within a radius of 15 km. If it was located on a raftable watercourse the supply area was extended by a strip of 30 km width along this water. All wood outside this area was inaccessible. 'The best wood, the densest forests, were without value if they could not be opened up by rafting. ... Therefore, there were wood deficiencies and price increases in many places, while only a few hours further wood was without value' (Wilsdorf 1960, 36). People lived in islands of scarcity. Average values for large areas were virtually irrelevant to them. This fact is very important in the consideration of preindustrial wood scarcity. It mattered not if there was still untouched forest somewhere in the interior. Only what could be obtained through a justifiable outlay in energy terms counted.

If the forest is considered in terms of fuel provision, it must be remembered that the transportation cost in making wood available for the consumer was part of the total energy outlay. Wood as an energy carrier was only useful if it could be obtained with justifiable transportation cost. This meant that in terms of energy wood was scarce if the energy investment in transportation was higher than the energy yield to the consumer. Total energy costs comprise the sum of the expenditure in wood (forest area), draught animals (pasture) and human labour (arable). The preindustrial energy budget may therefore be expressed in area equivalents since in a solar energy system the area, which stores photosynthetic energy, is critical. The only exception is kinetic energy (water, wind) which moves through the biosphere independent of its use. The wood losses that arise during rafting can be expressed as an area equivalent, so that with a possible loss of 25% of the rafted wood, it is not entirely unreasonable to establish coppices near settlements.

Until the 18th century transforming different energy forms into one another was only possible to a limited extent – a reason why no uniform energy concept could exist. Thus, it was not possible before the invention of the steam engine to transform the chemical energy of wood into kinetic energy; its use was bound to the release of thermal energy. Substitution consisted of alternative land uses: if woodland was cleared and turned into pasture, draught animals could be fed that supplied the desired mechanical energy. On the other hand, the mechanical work of a horse could be used to win thermal energy if it transported wood, but possibly the requirement for land might be higher than if one did without the horse and reforested the pasture. In essence, the issue was the optimisation of land use ratios. This was the key characteristic of the solar energy system: in the last instance, the amount of energy available depended on the area on which the sun's rays fall and where they are photosynthetically fixed. The management of the energy system consisted in optimising the relationship between different kinds of areas and the efficiency of their utilisation.

It may be asked why this optimising of area relations could not have been controlled by the pricing mechanism. Economic theory claims that a more efficient land use is expressed in a higher rent, so that land prices or rents could have signalled the appropriate alternatives. But this process would have required fully developed markets, expressed in complete alienability of land, a high mobility of capital and a certain homogeneity and transparency of markets. In preindustrial society these conditions did not exist. Instead, there were always qualitative aspects at work, such as seasonal availability of draught animals and the possibility of eating them in a situation of distress; but in particular legal and cultural obstacles stood in the way. Land was not just a factor of production, but a source of power and status. The pricing mechanism and rents could only have been conclusive with regard to energy relations if energy use had been the sole deciding factor.

2. Preindustrial Wood Consumption

In contrast to the peasant economy, the trades that inevitably rose with the late Middle Ages considered wood as a marketable good from the start: it had a price and had to be bought like other raw materials. It was the trades that generated the enormous demand that finally resulted in the wood crisis of the 18th century. Let us consider more closely in what quantity and for what purposes wood was consumed.

Commercial consumption

In the first place, the legendary demand of glass works for wood must be mentioned. For the production of 1 kg of glass, no less than 2,400 kg of wood were needed, 97% of it in the form of potash (Gleitsmann 1980, 116). Only 3% of the wood used was actually consumed generating heat, but that still amounted to 72 kg per 1 kg of glass. This shows that the demand for heat was high during glass smelting. It amounted to 90,000 kJ/kg glass with wood firing and the efficiency of fuel utilisation lay around 0.3% (Kahlert 1955; this figure only applies to the Middle Ages. Efficiency improved in the 18th century as a result of efforts to save wood). Of course, the ashes resulting from heat production were also used as a raw material, but it was by no means sufficient. The greater part of the potash had to be acquired separately. Glass blowers attempted to buy as much as possible from private households, but ashes were an important part of the farming and cottage craft economy. It was by no means a mere 'waste product' of heat generation, but was also used in the manufacture of soap and of alum, which served as a dye and a bleach. Therefore, glass works directly depended on burning potash from wood.

Given the enormous demand for potash, glass works were considered 'a wood-devouring thing' (Zedler 52, 1747, 1166). Thus, it was said that glass manufacturers in Silesia alone used roughly 500 tons of potash annually during the 18th century (Ganzenmüller 1956, 20). According to our calculation, this would have required a primeval forest stand of about 1,000 ha, or the sustainable yield of 100,000 ha of forest. A glass works was not entirely independent of other local conditions either, since apart from potash it also required silicon in the form of quartz, gravel or sand. Where that was present, glass making could follow the wood reserves. Such relocation 'according to the wood' was common in the preindustrial period. There are a series of examples of 'daughter huts' forming at some distance from original huts (Bloss 1977, 44ff.).

Salt works were another large-scale consumer of wood (Multhauf 1978). Since ancient times salt had been won by evaporating sea water in the sun. This was not really possible in the interior, or on the coasts of central and eastern Europe. For this reason graduation houses were set up: these were installations made of wattle that had a huge surface area so that water would be evaporated by ambient heat. On the continent the brine of saline springs was often evaporated in large pans over a flame. Wood consumption was enormous because of low efficiency. Depending on the concentration of the brine, 15 to 100 kg of salt could be produced with a cubic metre of wood. The Hall salt works in the Tyrol used a million cubic metres of wood for the production of 14,000 tons of salt in 1515, which is equivalent to the annual yield of a forested area of 200,000 ha – this is almost the entire Tyrolese mountain forest. (Rubner 1967, 32). This is an extreme figure because lower consumption figures are cited for other salt works: in Reichenhall the annual wood consumption fluctuated between 1520 and 1630 from 84,000 m³ to 210,000 m³, with a maximum of 250,000 m³ achieved in 1611. Hallstadt, Ischl and Ebensee in the Salzkammergut district and Aussee in Styria together consumed more than 150,000 m³ in 1524. In Halle, Saxony, it was annually 100,000 m³ and in Lunenburg about 150,000 m³ (Gleitsmann 1980, 114; Bülow 1962, 105). It can be rightly said that wood consumption by the Lunenburg salt works created the Lunenburg Heath (Wagner 1930).

A series of other trades dependent on wood can be listed. For consumption that was not energy-based, the building industry, which experienced an enormous expansion in the 12th and 13th centuries, takes first place. Even if the building itself was made of stone, as with Gothic cathedrals, roof construction required a large number of oak beams. Half-timbered housing also required much wood. Even when houses were made of brick, there was no getting by without fuel. All in all, more wood was consumed when a house was made of brick than if the whole house had been made of wood. However, that was only true if no peat was used for firing the bricks. Also, a brick house was more durable with fewer repairs expected than a wooden house, so that brick conserved wood over the long run.

All kinds of tools and machinery were made with a vast consumption of wood. Plenty of examples can be found in contemporary book illustrations: in looking at the illustrations in Georg Agricola's work *De re metallica* of 1556, one is astonished by the huge quantities of timber used in the mining industry. Not only are the tunnels supported by wooden columns, but the assorted machinery and waterworks required to produce the ore, the hammer mills and bellows, the carts and wharves for boats were all made of wood. The same applies to military fortifications and siege engines, so that disruptions during wars did not always permit forests to recover. War often devastated more wood than peace-time trade.

Other large-scale consumers were the boiling works for the production of alum and vitriol, the 'poison huts' for sulphur and arsenic production, potteries, lime kilns (lime was used not only in construction but also as a fertiliser), breweries, bakeries and tanneries, shipyards, the pitch and soap works, the manufacture of wine presses and barrels for viticulture, and generally the making of all sorts of containers and packaging such as boxes, baskets and tubs through to wagons and boats, sheds and fences.

Iron smelting

Large quantities of wood in the form of charcoal were used in smelting and processing metals. This was mainly obtained from hardwood, especially oak and beech. The most suitable were 20-year-old stems from coppice plantations. For carbonisation, trunks were reduced to segments one metre in length, then piled up conically around a post so that a kiln of 6–10 m diameter resulted. It was heaped with clay, then the post in the centre was removed and glowing charcoal was poured into the hole. Heating under airtight conditions broke up the carbon-hydrogen links in the wood, steam and gases evaporated and wood tar flowed out of the kiln. Almost pure carbon remained. After the wood had slowly carbonised and the charcoal had cooled, it was sorted according to size. Only large pieces were suitable for use in the furnaces, since only they left enough spaces to allow air to circulate through the admixture of ore and charcoal so that iron oxides were reduced and the charcoal oxidised.

For domestic use, brush and wood waste was often carbonised in the open or in a pit, during which the wood was sprinkled with water to prevent ignition. The size of these charcoal pieces did not matter. Finally, particular types of wood coal with particular chemical properties were produced, such as aspen wood for the production of gunpowder.

Charcoal for iron smelting could not be transported any distance overland, because the larger pieces would be pulverised on the poor roads. It has been calculated that overland transport of charcoal was only reasonable over distances of less than 7 km. The possible supply area for an iron works was thus clearly limited, but it should be remembered that such restrictions were obviated by transport over water.

To estimate the magnitude of wood consumption in an iron works and to obtain indicators for site conditions, the following calculations can be made (cf. Hammersley 1973): in early modern England 2.5 loads of charcoal, obtained from 800 cubic feet of wood, were required to smelt 1 ton of pig iron. To produce 1 ton of wrought iron, 3 loads of charcoal were required; this is the equivalent of 960 cubic feet of wood. It should be remembered that since losses occurred more than a ton of pig iron was necessary to produce a ton of wrought iron. To yield 1 ton of wrought iron from iron ore, approximately 1,800 cubic feet or 50 m^3 wood had to be used. Thus, sustainable production of 1 ton of wrought iron depended on the annual yield of a coppice plantation of 10 ha (cf. Rubner 1967, 34; Flinn 1967, 117).

Consider now that the maximum transportation distance for charcoal is 7 km and we can estimate the possible capacity of an iron works. The wood of a coppice with an area of 7 km radius or 15,000 ha could annually produce 1,500 tons wrought iron under the conditions described. Since the supply area increases by the square relative to the transport distance, annual production could rise in an extreme case to maybe 2,000 tons per site. Here wood supplies alone set an upper limit for the iron works. Even if wood instead of charcoal was transported to the iron works to be carbonised on site, the capacity could not have been increased given the enormous energy expenditure in transport.

Furthermore, there was another important energy barrier to the size of the iron works. The operation of bellows and forge hammers was dependent on mechanical energy. Only water power was suitable, since a horse gin was far too expensive, and given the required continuity of production windmills were inadvisable. A blast furnace using charcoal was in operation for days and weeks without interruption. There could be no disruptions of production due to a lull in wind. However, under the prevailing technical conditions mechanical energy could only be transported over short distances because costly transmission systems were needed for this purpose and friction soon became excessive. An iron works therefore had to be set up in the immediate proximity of running water with a sufficient gradient, limiting the number of possible sites by this condition alone. A blast furnace could not be located like a glass works merely where sufficient wood supplies existed: there were further conditions to be fulfilled in proximity to iron ore and adequate water supply. These requirements were most likely to be satisfied near rivers. This not only made the import of ore and the export of iron products by water possible, but could also enlarge the area of wood supply using rafting. Under favourable conditions it was possible to supply charcoal by water over a longer distance (Lindsay 1975).

In all these calculations it should be remembered that the data only give a crude orientation and wide deviations in either direction must be allowed for, depending on regional conditions. For this reason information on historical wood consumption varies substantially; not least of the problems being conversion of the measures given in the sources into modern values. In our model, calculations with approximate rather than exact values are used. The actual range was large.

Thus, the number of iron works in the Upper Palatinate rose from 77 to 200 between 1387 and 1464, their production increased from 2,000 tons to 10,000 tons and demand for wood from 175,000 m³ to 400,000 m³ (Lutz 1941). The average values clearly lie below the upper limit of 2,000 tons per site calculated by us. I know of no example where this figure was actually reached. In early 18th century England the technical upper limit for an iron works was considered to lie about 800 tons annually (Flinn 1958). This barrier was set by the size of the blast furnace. If it was too large, the weight of the charge crushed the charcoal in the lower reaches so that no draught occurred. Therefore, a single iron works could not utilise the maximum theoretical supply in one location. However, the concentration of several works in one location was not possible because of the implied demand for mechanical energy.

But the wood consumption of iron works still led to severe forest devastation. Wood demand for the Upper Palatinate iron works exceeded the overall increase of the forests there, so that in the opinion of one forest historian 'the Upper Palatinate forests, which are not very productive by nature, were virtually plundered by iron processing' (Rubner 1967, 31). In 1768 the iron industry in Carinthia consumed 300,000 cords of wood in 300 smelting works, i.e. around 700,000 m³. In the 16th century that had been only 220,000 m³ (Wilsdorf 1960, 11, 122). Such an increase in consumption soon had to reach the limits of annual growth, especially since different trades often competed with each other for fuel, and always with private households.

Private households

Consumption by rural and urban households played a role that should not be underestimated. In the year 1768, the 30,000 private hearths of Carinthia required over 1 million m³ of wood: that is more by half again than the Carinthian iron industry consumed in the same year (Wilsdorf 1960, 11). The construction of a single farmhouse in the Black Forest in the 18th century used 1,000 to 1,500 m³ of wood (Mitscherlich 1963, 10). Furthermore, there was demand for all sorts of sheds, tools, wagons, fences, pipes, spinning wheels, looms, furniture and so on. The consumption of firewood in particular was enormous. It is reported that the court in Weimar devoured 1,200 cords of firewood annually in the 16th century; in 1572 it was even supposedly 1,317 cords (Sombart 1919, II, 1140). If we estimate a cord at 2.3 m³ that would be 2,750 and 3,000 m³ wood, equivalent to the annual yield of 550 and 600 ha of coppice. In Villingen every household was entitled to firewood of up to 35 m³ annually; the firewood allowance for employees at the University of Königsberg in the year 1702 was no less than 70 m³ (Mitscherlich 1963, 9).

Energy efficiency was exceedingly low in private wood burning, probably under 10%. Rural houses in the Middle Ages were usually heated by an open hearth in the middle of the house. Meals were also prepared there. Smoke spread through the house and escaped outside through cracks and gaps in the ceiling and walls. The farmhouse did not have a chimney. One can imagine that such a house could not be too well insulated against heat loss if the inhabitants were not to suffocate in it. Soot and bad air were accepted because a smoky house also had advantages. 'The woodwork is protected against worm damage in such a house, the thatch lies much longer in a smoky house than in a house with chimney. Even vermin ... is supposedly kept away by the biting smoke' (Faber 1957, 24). Foodstuffs in the attic are also preserved. The smoke stench in houses was proverbial. A saying of the 11th century was:

> Sunt tria damna domus
> Imber mala femina fumus
>
> (There are three evils in a house:
> damp, a wicked wife and smoke)

To reduce the last evil, a number of attempts were made to improve smoke draught. In single-storey buildings an opening in the roof (*Windauge* = wind eye, window) was created through which smoke could escape. Later a flue that opened into a separate attic floor was placed in the room above the hearth. There have been chimneys since the 10th century. The fireplace was walled in on three sides and a stone chimney led into the open. This was not without problems, since a draught would arise and a thatched roof could easily catch fire from flying sparks. The chimney only gradually established itself in northern Europe, including in Germany, since a larger amount of heat escaped outside than from the open hearth. Demand for firewood rose drastically with the introduction of the chimney. In residences of the nobility it nevertheless became customary, since the amount of wood consumed was not as important there. Women's quarters in medieval castles obtained their name from the new comfortable installation (*Kemenate* = *Caminata*). Up to the 14th century the chimney remained restricted to the houses of the upper classes.

In multiple-storey urban houses with a number of tenants, heating with an open hearth was no longer possible. Here chimneys were built that were linked to a fireplace on every floor. Cast iron chimney plates that stored and deflected part of the heat back into the room were used from the 15th century. They first established themselves in England or France and from there also reached Germany.

In the countryside the tile oven was developed, especially in the colder parts of south Germany and eastern Europe. It probably owed its origin to the baking oven that had to be fired from the outside, gave off heat to the inside, was attached to the house and then extended into the room. It was later decorated with tiles and adopted by more refined households. The iron-plated oven was also heated from the outside at first (from the kitchen or the hallway). It established itself in the 17th century.

> From the point of view of firing technique, both chimneys and the old tile and iron plate ovens were quite incomplete... The smoke of the wall oven escaped through a door in the fire hatch or a hole above it, in the best case into a far too wide flue in the kitchen. The ovens devoured unimaginable amounts of firewood but only released a small part of its heating value into the room; usually four fifths escaped uselessly through the flue or stove pipe. Insofar as the early room stoves could be serviced from the outside and kept smoke-tight, they did not release fumes and smoke into the room, an inestimable advantage that was indirectly critical for an improved living culture (Faber 1957, 82).

The increase in living comfort was at first associated with an increase in the consumption of fuel. However, insulation measures such as panelling the walls increasingly became possible, since the room did not have to be continually aired. On farms, especially in Lower Saxony, the open hearth in the hall remained customary until the 20th century. Here the animals that lived under the same roof provided some warmth because the stable was open to the living area.

3. Regulation Problems

Today there is a widespread notion that the medieval forest was considered a wilderness and that wood was a free good like water and air for all to use as they pleased. From this perspective, it is reasonable to expect that the varied uses of the forest would soon lead to overexploitation. The forest without rules and laws would inevitably enter a fundamental crisis at some point, from which only the regulating intervention of the state could save it. However, the example of the forest can show that historical development was not a move from lawlessness to lawfulness, but merely from older local forms of regulation to newer territorial forms in the modern period.

In 1968 the American ecologist Garett Hardin presented a simple and appropriate model that was supposed to explain why unlimited access to free resources would quickly lead to ruination. Hardin illustrated this 'Tragedy of the Commons' with the village commons, a communal pasture. Since woodlands in Germany were largely communal property until the 19th century, Hardin's considerations may also apply to their use.

Hardin's model, which is based on the 'prisoner's dilemma' in game theory, has a simple structure presented here with a slight variation. Assume that a pasture of a given size has a maximum carrying capacity of 100 head of cattle and is the common property of 10 cattle breeders. In this situation each farmer could drive 10 beasts onto the pasture without overexploiting it. However, an individual farmer can make the following calculation: if I drive 11 instead of 10 cows onto the pasture, the entire herd grows from 100 to 101, and the carrying capacity is exceeded by 1%. Each of my cows only receives 99% of the fodder available to each of the 10 animals before. Therefore, my overall benefit has not risen by a full 10% but still by 8.9%, so I am acting rationally overall.

However, a similar consideration can be made by each of the participating farmers any number of times. Overgrazing of the common pasture is unavoidable since everybody can gain by driving in additional animals. As a result this maximisation of individual benefit will destroy the pasture so that in the end the benefit to each individual will disappear. Still, there is no easy escape from this dilemma for the individual. A farmer who predicts the long-term effect of this behaviour and so forgoes driving more than his sustainable 10 beasts onto the pasture harms only himself, because he rejects his own additional individual benefit but still shares in the growing collective damage. In the end, his restraint only gives increased scope to the expansionist behaviour of the others. Hardin draws the following conclusion from these considerations: 'Freedom in a commons brings ruin to all' (Hardin 1968, 1244).

What has been labelled the Tragedy of the Commons is a fundamental structural problem in every society. The individual benefit of a certain action is in many cases larger than the individual share in the collective damage that is associated with this act. Hardin depicts the case of a society of private proprietors in which each can use the common property, which may be exhausted, independently of the others. However, this is a special case. In the end it is true of every societal norm that personal benefit may accrue to transgressors while transgressing this norm does in fact cause them harm as members of society but to a lesser extent than their benefit. If all members of society made this rational individual calculation, the total damage and, therefore, the shared individual damage would be enormous. Real societies usually threaten sanctions in the case of a violation of standards so that a tangible direct negative effect in the form of a penalty, ostracism etc. augments the minimal indirect share in damage. Thus, societal damages become individually tangible and even the rational calculation of benefit conforms to socially acceptable actions.

In a normal case one might expect that societies would not permit a Tragedy of the Commons to occur. In particular, in the case of the common pasture and the forest it can be demonstrated that agrarian societies were capable of developing and enforcing rules that aided the prevention of such a process.

Traditional societies usually possessed a series of possibilities and methods with which to control socially harmful economic behaviour: restriction of use by collective consensus, which is enforced by social control or internalisation of behavioural norms; development of individual motivation that makes maximisation of individual utility less desirable; systems of symbolic redistribution by which the accumulation of private wealth beyond social obligations becomes impossible; patterns of preference that value leisure more highly than material wealth; direct threat and execution of punitive sanctions; and much more (see Ostrom 1990).

The agrarian order and organisational forms of medieval society were complex, and there was no clear directional tendency in its development. Still it is evident from the sources that village and rural communities had a high degree of regulatory power as far as the fields and forest terrain were concerned. This corporate organisation should not be misunderstood as a relic of a primitive communism, since it often had a seignorial past. It was not egalitarian either, but rather shaped by lordship. Two levels must be distinguished here: first, there were feudal relations between a liege lord and his vassals, the landlords, in other words a relationship within the nobility. However, on the local plane the relationships between peasants or peasant communities and landlords were significant; they were often complicated by the fact that a community could be under the influence of several (secular or clerical) lords.

Rules for the use of the commons came about as a result of negotiations between landlords and peasant communities. These regulations therefore had only a restricted spatial extent, so that the principle trait of agrarian society, its decentralisation, was also effective in the regulation of access to natural resources. Rural communes regulating the commons were closed communities which made a precise distinction between members and non-members, between participants and non-participants and whose members possessed socially graded rights. Important institutions were communal assemblies and wood courts (Holthinge), whose results were set down in special legal acts, the Weistümer. These articles of corporate law demonstrate how resource management was executed and what regulations were attached to individual farms.

In the early Middle Ages (a period for which written sources are scanty) taking wood in the communal forest was hardly limited. Each member was entitled to as much for construction, tool making and firewood as required for personal economy. However, there were almost always restrictions on clear cutting and to some extent on removing fruit-bearing trees, especially oak, which was important for pig forage. The use of the forest as pasture and for pig forage (pannage) was often regulated. Frequently, keeping goats in the forest was forbidden since they seriously harmed growth and regeneration. When wood or other products, such as grass, small game or fish, were taken unlawfully, the offence was punished. This was certainly the case with removal of planted wood, masting trees and felled wood.

The penalties threatened for transgressing the forest regulations were often draconian. A *Holting* of the Deisterwald decided in 1528 on the question of how someone should be punished who debarked oak and fruit trees (*eckbohme oder andere fruchtbohme schelde*). This was a widespread practice to cause trees to die off to clear sheep pasture.

> Resolved: if one does this and knows better, he shall be cut open and his entrails shall be pulled out of him and tied around the tree (Grimm 1957, vol. 4, 669).

A wood court in the Hülsede Mark decreed that if anyone lopped the crown of a fruit-bearing oak tree, his head should be planted in its place (Timm 1960, 70). Clearly these penalties contained symbolic elements in the spirit of the *jus talionis*, i.e. a retributional penalty. Human entrails were to bandage the debarked tree; the human head was to replace the crown of the tree. It is questionable if these severe punishments were ever really enacted , or whether a monetary fine was deemed sufficient in any actual case.

Apart from communally regulated forests, there were woods at a distance from settlements that were not regularly used by anyone . Supreme ownership of these was claimed in the Middle Ages by the kings, as part of the royal or imperial demesne. This right was later proclaimed by a formal act of inforestation. The crown land was withdrawn from general use, it stood outside (*foris*) access by village communities – it became Forest. Hunting, clearance and settlement in this territory were forbidden, and also all forms of agricultural use without explicit permission. With the expansion of sovereign authority from the High Middle Ages, the territorial lords claimed this right of inforestation for communal forests and the few separate private lands as well. It first affected the Royal Chase, the exclusive right of the sovereign to hunt large game.

From the 16th century onwards the trend was set towards a general super-regional regulation of forest use by higher authority. The princely right of inforestation was interpreted as the sovereign right to the forest and hunt. Insofar as this right claimed a fiscal right of material usage and occupation of items that were not fiscal property, they were generally classified as 'regalia' (forest and hunting regalia, but also mountain regalia etc.). A complete enforcement of these regalia was not always possible but had to be effected in agreement with the estates where they participated in governance.

The situation in the modern period until the 18th and early 19th centuries resembled a permanent tug-of-war between sovereign and local forces over the authority to regulate forest usage. In the context of Roman Law the forest regalia were generally considered as proprietary rights of the prince to forests, while the rights of village communities were merely considered material entitlements or servitude. In practice this did not change much at first, but eventually the tendency was towards pressing community members to mere entitlements of usage and to treating their rights of usage as revocable favours, so that the forest could be considered as the exclusive property of the prince or the state.

A whole flood of princely regulations in the early modern period were derived from the claimed lordship of the forest and were presented as protection of the forest, but in the end served fiscal purposes. The forest historian Bernhardt cites a series of sch regulations:

> The right to appoint higher and lower officers of the forest, to erect forest and hunting lodges, to enact forest and hunting jurisdiction, to command and forbid in hunting matters, to regulate all timber, to permit herding and pasturing, to command and forbid charcoal works, to permit mowing of grass, to use the mast, to exercise the taking of bees and fowling or to transfer [these rights] to others, to burn ashes and to establish glass works, also [to regulate] the admission of unskilled marksmen to the hunt, to forbid debarking trees and all clearances; furthermore the right to demand hunting socage and the delivery of all found stag antlers, to collect the clearance tithe and forest taxes, to tend arable and meadows planted with wood, to order clogging of dogs, to demand food and drink for the foresters, dogs and horses on farms, to demand the forest and dog oats, to set a time for the wood cut, to bind over forest malefactors, to set wood measures and prices, to command that subjects did not have the right to forestall any wood sales, to forbid pointed fences, to regulate wood days for collecting wood by hook and crook, for capping trees, pitch making and resin collecting, to forbid carpentry in the woods, to establish and regulate wood rafting, to command planting of trees, to appoint wood cutters and give them regulations, to regulate wood markets, to inspect the agriculture of subjects, raking moss, stripping leaves, cutting rods and broom and other damages, also to forbid common people to bear the hunting knife and wear the green forester's cloth (Bernhardt 1, 1872, 230).

These regulations affected not only the prince's own forests but also woodlands in the possession of communities and corporations. The disposition of property was clearly limited by these regulations of higher authority and the competence of wood courts was reduced. The concrete occasion for this process was that wood and land became more scarce as a consequence of denser settlement, so that prices rose with increasing demand. The increasing penetration of subsistence economy by the market and the novel luxury goods imported from overseas caused the princes' requirement for money to rise, so that they saw a means to boosting their income in the appropriation of forests. The growth of towns created a market for wood. It acquired a value and its ownership began to be disputed.

In the face of heavy resistance by those affected it is unsurprising that not all communal forests could be entirely transformed into the private property of landlords or princely sovereigns. They often persisted until into the 19th century and even today in Germany approximately a quarter of the woodland belongs to the communes. However, in the course of modern development it became possible for commons to be divided into private property on petition by some community members. This tendency was somewhat buffered by the transformation of the communes into public and legal organs of the state, through which they acquired certain duties such as the organisation of care for the poor, which they in turn financed through the revenue of the communal forest.

The most important consequence of the enforcement of territorial rule in the course of the modern period was the extensive displacement of communal woodland regulations by state forest regulations. These were frequently formulated with co-operation from the estates, i.e. the local landlords, and regulated by general laws the use of land and management of all woods within the territory, regardless of whether they were princely, corporate or private. Some forest regulations were already in place in some areas in the 12th century but the majority only appeared from the 16th century onwards.

Almost all forest regulations by higher authority began with a complaint about the increasing devastation of woods and impending scarcity of wood. Thus, the forest regulation of the Bavarian Duke Albrecht V of 1568 stated:

> If one were not to intervene now, all our subjects and residents, rich and poor, in towns, market-towns and in the country, would generally encounter, and there would appear within a most short period of time, a noticeable and burdensome scarcity of wood (with which one cannot do without), the which to acquire many of them would leave their goods, homes and sustenance including even their wives and children and go from the same because of its lack, and irretrievable damage would result from it (Köstler 1934, 101).

Likewise, the forest and wood regulation of the Elector Augustus of Saxony stated in 1560:

> ...that Our forests and woodlots would go further into such decline, as has obviously already happened in several and part of most, We have for several weighty causes, such as a lack of wood for mines, to preserve Our game and to leave a not insignificant amount of wood for our descendants to have the comfort of building and firewood, but also that necessity requires that a part of Our woods and woodlots be closed and barred (Schmid 1839, 3f.).

Such invocations of a future lack of wood clearly bear legitimising traits. They can be interpreted in the sense that the looming scarcity of wood was the occasion to establish a state forest regulation and, therefore, justify a further expansion of the reach of the sovereign law and the absolutist state. The tendency of authority to expand into a legislating territorial state and to draw competence to itself doubtless existed independently of the scarcity of wood and the forest regulations that reacted to it. Thus, the perception by the state of a looming scarcity of wood signified a wish to monopolise governing powers. However, it cannot be deduced from this that there was no real tendency towards a scarcity of wood. On the contrary: if it was to serve as the means of legitimisation for expanded state regulation, it had to be apparent. A reference to a wood shortage as a result of which the reduction of traditional rights had become necessary could not be convincing in the face of an apparent surplus of wood.

Today we may ask whether the establishment of forest regulation by the state was necessary for the preservation of the forest, or whether traditional self-regulation by village communities would have been able to assume this task by itself. In reality there never was such an alternative. The village communities were undermined and weakened when wood was still in relative abundance. Their regulating function no longer had any force by the time wood became scarce; the communal organisation had already been disempowered for the benefit of the state.

Perhaps the overall movement can be summarised as follows: in the waning Middle Ages rural community bodies existed that had a high degree of competence in self-regulation because the regulating impositions of the kings were few. Wood courts were in a position to look after sustainable use of woods. On the way to the formation of the modern state a legal administration developed that went about expanding its competence into all possible spheres of society. In the course of this process special interests and privileges, local self-administration and corporations were increasingly subsumed into a general law that first appeared in the guise of absolutist rule. State forest legislation now took the place of communal wood regulations.

This process was considered a disempowerment by those affected and they attempted to stand up for their traditional 'rights and freedoms' (cf. Blickle 1986). Altogether this process, which may be understood as the establishment of the state's monopoly of force, was unstoppable. However, it constituted an important political background for enforcing forest regulations: these were not, as the older forest historiography viewed it, the altruistic exertions of the state to 'save the forests', which were threatened by the blind and shortsighted self-interests of the peasantry. The principle of sustainability was not an invention of forestry sciences in the 18th century but the general foundation of the peasant economy to which the use of forests belonged. Objections to this view which point to the power implications of forest regulation and warn against taking every hidebound complaint about scarcity of wood at face value are justified (Radkau 1983; 1986).

Let us return once more to the general problem of the use of common goods. A crisis-like peaking of the Tragedy of the Commons in Hardin's sense may perhaps only occur when a novel social, economic and technical pattern forms, that is, when the scope of action for individuals increases. The possibility of collective damage arises only when new elements of the natural environment are exploited and this exploitation is felt as a destruction of resources. With population growth and an increase of energy and material flows, ever more components of natural systems, which can only bear human activities up to a certain point, suffer more severely than before, and in this context a new need for regulation may arise.

Thus, running waters can incorporate the sewage of smaller settlements without affecting water quality downstream. However, when immissions increase in volume, water will eventually become a scarce good that must be husbanded. It will be sensible to establish binding rules regarding the inflow of sewage. Since this is associated with costs, those affected will first resist these infractions of their hereditary rights. If they are successful the problem analysed by Hardin occurs: everybody pollutes water and in the end nobody has clean water.

The most elegant solution to the commons problem is without a doubt privatisation. Nobody will pour sewage into his own well. If the good to be used is completely at the disposal of one individual, that person will bear the entire benefit and damage. If he calculates rationally, the damage will never exceed the benefit, as may be the case with common goods. When the proprietor acts in rational, enlightened self-interest, excessive use will be avoided.

Although privatisation may be elegant, it can never be realised completely. Individual natural components cannot be totally isolated from each other in such a way that actions in one area do not give rise to consequences that affect others. In practice, privatisation is incomplete in almost all cases; also, common goods always are made to bear consequences, which is why the use of private goods always contains parasitical elements. Privatisation is only reasonable as a solution to the commons problem if its external effects are minimal so that the common good can buffer them without being seriously harmed. However, after a certain size or density has been reached, this is no longer possible.

The commons problem may also arise when the social regulating system changes its character and function so that structural deficits, in a sense areas without rules, arise. This would be the expression of a normative pioneer situation. Generally, the commons problem may be understood as the expression of a need for regulation that comes about as an extension of the range of action but without it being possible to find new standards or mechanisms for their enforcement within a short period of time because this would conflict with traditional 'freedoms'.

This problem is easily recognisable in peasant resistance to the princely forest regulations. From the peasants perspective every limitation of use was an expression of princely arbitrariness disguised as impending wood crisis. To some extent they were right in this observation, but only in part. That they justified their resistance against the authorities with the argument that wood was an inexhaustible resource is evident, for example, in a source from Brunswick of 1650:

> Thus, We must treat with unbending strictness the resistance [against the pasturing prohibition in the forest – RPS] and also, furthermore, that some rough snot noses, when they are punished by Ours for this, to Our noticeable disgrace and to the greatest damage and disadvantage to Us and Our own and all posterity loudly proclaim the following words 'Wood and plague will grow any day' (Schwappach vol. 1, 1886, 345).

The invocation of posterity by the officials of the absolutist police state was presumably not taken seriously by the peasants, for whom pasture was an ancient right in the formerly common woods now subjected to state regulation. The displacement of this right, even if it was associated with material compensation, as often occurred, threatened to destroy the peasants' domestic economy, of which forest pasturage had been part since time immemorial. Forbidding this traditional form of pasturage appeared all the more presumptuous since at that very time the stock of game in the forest rose markedly. The strictly guarded hunt of the princes, whose greatest pleasure was hunting spectacles, was propped up by heavy penalties, and its consequence was the multiplication of assiduously tended game. The stock was too great and damaged tree plantations as much, if not more, than cattle driven into the forest for pasturing.

Here a contradiction became apparent between the prince as a private person and a hunter on the one hand and a representative of the welfare and police state that was responsible for a well-ordered economy of the country on the other. This conflict was far too often resolved in favour of the private interest, the hunt. The transfer of forest administration to the master of the hunt did not encourage a resolution in the interest of forestry. Only fiscal and cameral interest in the increase of the raw material wood, associated with a strengthening and growing independence of the absolutist or state bureaucracy, finally brought about in the 19th century an increased concern for keeping game stock within reasonable limits.

At the time of the 18th century 'wood crisis', the utilisation of the forest was an area of complex tensions. On one hand, the agrarian solar energy system increasingly encountered ecological limits in the course of population growth and the expansion of commercial production, which was felt as a more intense conflict over natural resources. At the same time legal and economic organisation changed: the precarious balance of communities, landlords and princely rule tended to be replaced by a new dualism between the private economy of the individual and general state legislation that came about in the course of the 'division of the commons'.

The privatisation of communal land (in the English context: the 'enclosure of the commons') was associated with numerous social, economic and also technical problems because it had to go hand-in-hand with a spatial and functional segregation of individual economy. The peasant economy of the Middle Ages and the early modern period mainly relied upon the three-field system with collective forms of rotation (*Flurzwang*) and agriculture from which individual economies could not be easily separated. Privatisation of the commons required extensive transformation of the village economy. The rural underclass lost its basis for subsistence; land had to be resurveyed and paths had to be laid out; risk of harvest failures now had to be buffered by the households themselves (insurance by taking credit); instead of pasturing in the forest, animals had to be kept and fed in stables.

The new forms of forest utilisation were thus elements of a comprehensive reorientation. Access to the woods was controlled and restricted to a greater degree by the authorities, while at the same time societal control mechanisms experienced a decisive transformation. In the place of local self-government that operated on the principles of proximity and social control came organs of sovereignty, as the state in a society of private property owners developed into an institution that enforced the common interest (no matter how fictive) against the individual interests of these private property owners.

Now that common property was no longer protected against overuse by public social control on a local basis, the state could step in with the claim of protecting the common good against the transgressions of individuals. The enforced law of the state took the place of corporate self-government. Therefore, the regulation crisis of the forest was a characteristic of a transition period in which older forms of self-regulation no longer worked, while the new regulation of the legal and administrative state were already present in theory but had not established itself in practice. Forest use was adapted to the new economic and legal circumstances only when more efficient repression by the state under the rule of law made it possible actually to enforce forest regulations.

This historical process, which did not take place in Germany until the 19th century, depended upon a series of ancillary conditions. The complete transformation of the forest into a controlled site of exclusive wood production was associated with a functional decoupling of agriculture and forestry, with the displacement of traditional rights of usufruct by peasants, which in the older perspective were inseparable from the peasant economy. Only the transition to the fossil energy system enabled this transformation, because it lifted the intensive exploitative pressure of the 18th century from the forests. This dampened the struggles for these resources and that was probably a factor in increasing the acceptance of new state regulation.

III

England: Coal in the Industrial Revolution

1. Substitution of Wood by Coal

The epoch-making transition to the use of fossil fuels, which coincides with the Industrial Revolution, originated in England. The homeland of industrialisation was also the home of coal technology: it was in the British Isles that the production and combustion of coal on a grand scale began and here that the huge global transformation of the socio-metabolic process set in, which is still under way to this day. The use of coal chronologically preceded the actual Industrial Revolution; the commercial development which led up to this enormous transformation and which then took the now familiar explosive course was directly linked to the consumption and growing range of applications for coal. The revolutionary acceleration of industrial development from the end of the 18th century is immediately related to two decisive technical breakthroughs in the use of the new energy carrier: iron smelting with coke, and the transformation of chemical energy into mechanical energy by the steam engine.

Why was coal first used in England?

The use of fossil energy constitutes the energy basis of the Industrial Revolution; this would not have been possible without coal. Why did England become the point of origin in this development? Was the occurrence and availability of coal the sole cause or did a series of other factors contribute to a specific combination which initiated industrial development?

Coal was already known in antiquity. Excavations in England but also on the continent, for example in the Saarland and the coal territory near Aachen and Liège, have shown that the Romans already used coal, probably for forging metals. Apart from such archaeological proof, there is philological evidence that coal was not unknown to the Greeks and Romans. However, it played no significant role in that period: apparently deposits were too much on the periphery of these societies to have a serious influence on fuel choice.

China, too, used coal at an early stage, no later than the 13th century, as is evident from the reports of Marco Polo and from Chinese sources (Hartwell 1962, 1967; Needham 1971, 135). Like so many other technological beginnings, the use of coal was not pursued and no commercial or industrial dynamic unfolded in China. The use of coal remained sporadic and locally restricted. In China, as in Europe, coal was used not only as a fuel but also as a raw material in various trades. Jewellery, boxes and inkwells were made from it and it was used as an interior plaster and a roofing material. The very hard pieces of coal used for that purpose were presumably found on the seashore and came from natural wash outs. The customary word for coal in England during the Middle Ages and the early modern period, 'sea coal' (as opposed to 'coal', which was charcoal), probably refers to this origin (Nef 1932, 2, 452ff.).

It is not known if there was a continuous tradition of coal use in England from the Roman period to the Middle Ages. Nor should one have any grand illusions about the technical sophistication of early coal use. In the beginning pieces of coal were collected when they came to light on river shores and slopes and were burnt for purposes if they could replace wood without disadvantage, such as work in the open requiring large amounts of process heat. Where this was possible, the beds were followed some way into the ground, but at most holes or pits were dug for this purpose. Actual coal mining underground did not exist either in antiquity or in the Middle Ages.

It is clear that only small amounts of coal could be produced in this manner. Its use was regionally restricted. Even in England wood remained the only fuel worth mentioning during the Middle Ages. Coal played no role in quantitative terms. England, like other parts of Europe, was densely forested and even after the clearances of the Middle Ages had sufficient stands of wood to secure the provision of the rural population with fuel (Darby 1950; Darby 1951; Smith 1978). Here, as on the continent, the usual scarcity of wood in densely populated centres existed but it was not caused by a real lack of wood. The problem was rather the great difficulty of transporting wood. Thus, on a small scale it might have been easier for a village smith to acquire coal from a deposit in immediate proximity than to fell trees at a great distance and to carry them with great difficulty to their destination. Even if forests were there in abundance, it was advantageous to use the higher energy concentration of coal. In this case the substitution of coal for wood may be attributed to advantages in transportation and not in a strict sense to a scarcity of wood.

A similar process took place on a grander scale when population centres had to be supplied with fuel. Here geographical factors were decisive. On the continent it was always easy for a location on a raftable river to obtain sufficient quantities of wood. Mostly, rivers were quite long and originated in mountainous, wooded areas. In the source regions of these rivers the land was often unsuited to agriculture because of the slope, the soil or the climate; these forests were only cut very late, if at all, because agriculture was hardly worthwhile. The middle ranges as well as the marshy flood plains were ideal locations: these areas rarely had to compete with other forms of land use and the rivers offered natural transportation routes for wood. They linked territories that were primarily used as woodlands because of their ecological disposition with centres of population, where people depended on a supply of wood.

This regional division of labour between woodlands and population or commercial centres became possible because the use of water routes suited the specific energy characteristics of transport on running water. Basically, little labour was expended in transport by water since there was little need to overcome friction. Also, no complicated transportation systems had to be established and maintained to reduce the investment of labour. Only pack horses or horses and wagons were options apart from the water route at that time. Pack horses had the advantage that they could get ahead even with very bad roads and weather conditions, but their useful load was far less than with wagon transport. If draught animals were used, only the rolling friction of the wagon had to be overcome, but a level road was required. The construction and maintenance of such roads was only worthwhile if goods were transported on it regularly and in great quantity.

By contrast, the waterway offered many advantages. Basically the boat was the ideal means of transportation: on a water surface it only had to be moved on a horizontal plane. However, there was a significant difference depending on whether goods had to be carried upstream or downstream. Human and animal labour had to be expended for transport against the current. Still, towing was as a rule more effective and less costly than axle transport. Also, a towpath required less investment for its maintenance than a road on which heavy wagons moved. It was no accident in agrarian societies, in which energy was always in short supply and the most energy efficient process was naturally favoured, that towing systems and even canals were maintained for the transport of heavy goods. However, the effort of building a canal was so great, especially if larger differences in elevation had to be overcome and locks had to be built, that it was only worthwhile for routes on which large quantities of heavy goods were transported. Canals and navigable rivers were the basis of transportation of goods in preindustrial society, while roads were barely suited for heavy transport over larger distances.

By its nature, downstream transport was particularly easy because the expenditure of labour was particularly low. Rafts required only a small crew, mainly occupied with steering. A regional division of labour was indicated by the wood procurement process: trunks were easily floated down from the upper reaches of a stream, high value goods were transported upstream. This structure of wood supply on the water route, which was determined by energy, favoured a complementary development: raw material production upstream, population and commercial centres downstream (cf. Cottrell 1955). Since a forest economy rather than agriculture was suited to flood plains and the mountainous upper reaches of rivers, this form of land use was well adapted.

This regional division of labour conditioned by energy could hardly develop in England's geographical circumstances. Rivers here are relatively short, have only a slight incline and their source areas are hardly in the continental sense mountainous regions requiring a forest rather than a pasture economy. The drainage area of rivers was not large enough to supply population centres with large quantities of wood. Thus, there were natural barriers to supplying London with firewood by the Thames, while the town of Rotterdam could be supplied by the Rhine with wood from the Black Forest. This explains why local scarcity of wood must have been felt more urgently in England than on the Continent.

This unfavourable supply situation of English towns with regard to firewood was in contrast to the ease with which they could be reached by sea. Transport by sea has fundamentally different characteristics from transport on running water and canals. In the first place there is no asymmetry in the direction of the flow. In principle, transport in all directions is possible with the same effort. There can be exceptions due to seasonal storms or prevailing winds; but the wind is an important source of energy that can be freely used like the current of a river. Once the navigation technology had been mastered and it was known how to cross against the wind, wind energy could be used without expenditure proportionate to the amount of energy involved. If a sailing ship is used, distances between places are hardly significant. While on land or during towing a similar amount of energy must be used for every segment of the distance, there is no direct relationship between distance and labour in sailing: once a ship has been built and has a crew, the distance it must cover is hardly significant. The relationship between effort and yield is no longer linear.

For this reason it was apparently more effective to ship coal from the region around Newcastle to London than to carry wood, even though it was far closer in the Weald of Sussex or a similar larger forest. It was as easy to supply London with coal as it was difficult to transport wood there. Therefore, a specific geographic situation favoured a massive shift to coal. The difficulty of firewood provision was juxtaposed by the ease with which coal could be transported.

In England two factors permitted the spread of coal utilisation beyond the local consumption customary on the continent: the occurrence of easily accessible and easily mined deposits in combination with a geographic location that facilitated the transport of coal while hindering that of wood. When we view other parts of Europe to see if there was a similarly favourable combination of factors elsewhere, we must come to the conclusion that the situation in England was unique. There were comparable deposits on the Ruhr, the Saar and near Aachen that were mined where they cropped up; their use remained locally restricted until the 18th century when people faced a more serious fuel problem. Possibilities for transport similar to those in England did not exist, nor was the supply structure of wood similarly bad. However, there was no coal in areas like Italy or Greece, where one could expect, given the state of transportation, that larger distances in all directions could be overcome by sea, and where wood was equally scarce. If it was to happen anywhere in Europe, the large scale transition to coal could only be achieved in England.

The rise and decline of coal consumption in the Middle Ages

If one considers these natural conditions, it is not surprising that coal was consumed in London soon after the onset of production on the Tyne in northern England. Already in 1228 the name 'Sacolas Lane' (i. e. Sea Coals Lane) is documented in a London suburb (Galloway 1882, 5). The geographical circumstances apparently so much favoured the use of coal that it began to compete with wood as a fuel at an early point in time. Its consumption increased step by step with the increasing demand for fuel. This substitution of coal for wood was at first sight an unproblematic process mediated by the market: not an absolute scarcity of wood but the lower price of coal effected its dissemination.

Among the first testimonials of coal use are complaints about noxious smells. Unlike the familiar smell of smoke from a wood fire, the penetrating smell of coal was offensive.

> The smoke of sea coal fires was a general nuisance in London by the last quarter of the 13th century. A royal commission appointed in 1285 to inquire into the operation of certain limekilns found that where the lime used to be burnt with wood, it was now burnt with sea-coal. 'Consequently, the air is infected and corrupted to the peril of those frequenting ... and dwelling in those parts' (Te Brake 1975, 339).

Repeated complaints because of the stench resulted in the frequent cry that coal use should be entirely banned in London. These efforts were successful when a general prohibition of lime burning with coal in London was issued in 1307 under Edward I, but apparently it was not taken seriously.

William Te Brake notes that complaints about coal stench abruptly stopped in the last quarter of the 14th century and only set in again in the 16th century. The reason could be a simple gap in documentation but in his opinion there are clear indications there were no more complaints about coal stench because the problem decreased. The use of coal sharply declined in London between the 14th and 16th centuries, even though it did not altogether stop – that is the most plausible explanation for the interruption in complaints. The reason for this decline in the use of coal is surely not to be found in royal prohibitions. There is no clear indicator that a sustained attempt on a grander scale was made to suppress coal burning. Such a policy would have been documented, if only in testimonials of resistance against the prohibitions. One might also expect that there would have been transgressions and punishments. Finally, lifting the prohibition should have been the subject of demands and campaigns that were in the end successful. The renewed rise of coal use in the 16th century, which is among other matters documented with a series of objections against this aesthetic irritant, was not preceded by a campaign against the prohibition.

We must look further afield for an explanation of why people turned to an increasing extent to coal instead of wood as a fuel up to the beginning of the 14th century, but then gave up using this apparently cheaper wood substitute for almost two centuries, only to resume its use again in the 16th century. Following Wilkinson and Te Brake, one suspects that the cause is to be found in a disruption of population growth. Let us look at the following figures for central and western Europe:

	650	1000	1340	1450
France, Low Countries	3	6	19	12
Great Britain	0.5	2	5	3
Germany, Scandinavia	2	4	11.5	7.5
All Central and Western Europe	**5.5**	**12**	**35.5**	**22.5**

Table 6. Population development in Central and Western Europe
(Millions. After Russell 1972, 36)

The population of Europe almost tripled between 1000 and 1340 AD. Of course, this meant tremendous pressure on the means of subsistence, especially in areas where it was not possible to turn little-used land into arable. On the continent the expansive settlement movement known as the Colonisation of the East took place beyond the Elbe. Also, woods were cleared in all other regions and the arable was increased that way, but in Germany about a third of the country remained forested. In England the forest decreased significantly – an important reason for the necessity of using an alternative fuel.

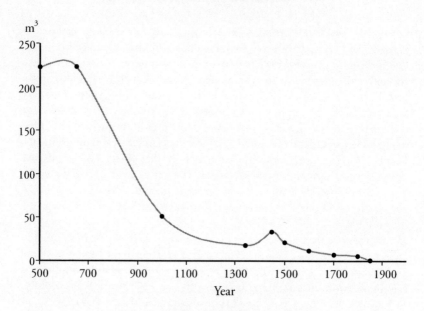

Figure 2. Fuelwood per capita in Britain

(This figure shows the hypothetical amount of fuelwood per capita in Britain AD 500–1850. The complete area of 228,000 km² is divided by the population, and 1 ha land per person is subtracted for non-fuel purposes.)

Medieval population growth was not accompanied by a significant increase in agricultural productivity. Mainly cereals (rye, spelt, wheat, barley, oats) were grown, also some vegetables and trade plants such as flax, hemp and madder. The heavy ploughshare had established itself in the early Middle Ages but the harvest was still cut with a sickle. Overall it was true that on the continent the state of agricultural technology in the 12th and 13th centuries was not significantly surpassed until the 18th century, except in the Low Countries, while England introduced the first significant improvements beginning in the early 17th century (Kerridge 1967). During the Middle Ages the net yield of grain was only two to three times the amount of the seed in all parts of western and central Europe. If one considers that weather-induced fluctuations of harvests accounted for about 20–40%, it becomes apparent that the available amounts of grain could be halved in a bad year. (Henning 1979, 79ff.). Since there were tight limits to storage through moisture and pests, and possibilities of transportation for bulk goods over larger distances was also limited, greater famines could result from harvest failures. There are indications that climatic conditions deteriorated in the 14th century: it became colder and moister (Le Roy Ladurie 1971). This worsened the situation and may have had drastic consequences for human health (Matossian 1986).

In the first place, peasants attempted to counter this problem by quantitative expansion, by increasing the arable and extending the exploitation of forests. In the long run this had to result in yield reduction. Soils less suited to grain were increasingly cultivated and there was a tendency to overexploit forests. This intensified the problem of looming harvest failure because the safety margin against climatic fluctuations diminished. Attempts to intensify agriculture, as they were developed in the 17th and 18th centuries in response to a similar mounting subsistence crisis, were rarely ever observed in the Middle Ages. Unrestrained population growth coupled with an inability to implement agrarian innovations finally resulted in an agrarian crisis, which was only resolved by the collapse of population in the wake of the Black Death.

As our table shows, the population of central and western Europe dropped by about a third between 1340 and 1450. Epidemics that swept in four waves across Europe between 1347 and 1383 were the immediate cause. It is an unresolved question whether this was a purely exogenous catastrophe or whether the poor nutritional state of the population may have been a precondition for mass deaths (see Herlihy 1997). In the first half of the 14th century famines and animal murrains increased. The amount of meat in the diet had declined significantly. It appears to be a reasonable conclusion that the Black Death was favoured by a poor nutritional situation. This is supported by the fact that population losses in areas with better meat supplies were limited. However, it is also proven that well-nourished bourgeois and nobles were not spared in affected areas. A single-factor explanation of the Black Death through a worsening of the nutritional situation is not without its problems (Wilkinson 1973).

However, McNeill (1976) sees an autonomous epidemiological process at work in the appearance of the plague. Thus, the epidemic is an exogenous incursion of a microbial parasite that could acquire the traits of a pandemic because of the high population density and the weak immune system of the European population due to a lack of protein. The wide distribution of the disease caused even better nourished individuals to be infected on a grand scale. Since the acquisition of genetic resistance takes at least seven generations, the plague was able to maintain its horror for some time. The subsistence crisis was a favouring factor in the success of the agent but not the immediate cause of the pandemic.

England was particularly hard hit by the Black Death. The population had shrunk to about 2.1 million after 1400 – a decline of 60%! In 1377 only 35,000 persons lived in London, as opposed to around 60,000 in the middle of the century. In the wake of the population decline the pressure on the means of subsistence eased rapidly. Unlike the Thirty Years War, which caused similar population losses in some regions of Germany, the plague only affected people, not livestock. Therefore, the nutritional situation suddenly improved. Meat consumption per capita tripled. The use of forests declined, less fertile ground was given up and converted into pasture if it was not abandoned to spontaneous reforestation by natural seeding. *Wüstungen*, formerly cultivated and then deserted settlements, were able to recover (Abel 1955; Abel 1978).

In this situation London apparently distanced itself from the use of coal. There was no need to use the unpopular wood substitute when the reduced population was sufficiently supplied with firewood. The demand for wood sank in absolute terms while its stock increased. The documentary gap for the use of coal in London can be explained by the improved supply of wood.

A new rise

The British population again grew rapidly after the Black Death. In 1500 it was already 4.4 million, by 1600 6.8 million and by 1700 about 9.3 millions (Mols 1974, 38). Due to commercial development the per capita use of fuels probably rose, while at the same time the availability of firewood declined because land was being used for other purposes. As a result the demand for firewood increased more than that of other goods, which was expressed in the fact that prices generally did not rise in proportion to its price. The following table, though it relies on very incomplete and, therefore, rather unreliable material, shows that the price of firewood rose tenfold between 1500 and 1700, while the general price index only increased by a factor of five.

	Price index	Price of firewood	Firewood/Price index
1501–1510	95	100	1.05
1551–1560	215	265	1.23
1593–1602	367	451	1.23
1653–1662	477	1026	2.15
1693–1702	525	1058	2.01

Table 7. Price Indices in England

(1451–1500 = 100. See Wilkinson 1973, 114)

Faced with this more than proportional price increase, coal was rekindled as a cheap wood substitute. The 15th century retrospectively appears as an age of abundance in natural resources, including wood. The scarcity of the age before the incursion of the plague was forgotten. It was believed that 'formerly' there had 'always' been a superabundance of wood. Thus, in an posthumous edition of John Stowe's *Annales or Generall Chronicle of England* it was said:

> Such hath been the plenty of wood in England for all uses that within man's memory it was held impossible to have any want of wood in England. But contrary to former imaginations such hath bene the great expense of timber for navigation; with infinite increase of building of houses, with the great expense of wood to make household furniture, casks, and other vessels not to be numbered, and of carts, waggons, and coaches; besides the extreme waste of wood in making iron, burning of brick and tiles. ... At this present through the great consuming of wood as aforesaid, and the neglect of planting of woods, there is so great a scarcity of wood throughout the whole kingdom, that not only the City of London, all haven towns and in very many parts within the land, the inhabitants in general are constrained to make their fires of sea-coal, or pit-coal, even in the chambers of honourable personages; and through necessity, which is the mother of all arts, they have of very late years devised the making of iron*, the making of all sorts of glass and burning of bricks with sea-coal or pit-coal (Stow 1632, 1025).

Obviously the memory of the time before the Black Death, when the use of coal in London was widespread, had paled in the meantime. A clear indicator of the renewed use of coal is the return of a multitude of complaints against noxious smells. Te Brake cites a series of voices from the second half of the 16th century that express a similar distaste to the 13th century for the novel custom of using foul-smelling coal instead of conventional wood (Te Brake 1975, 356). Rueful reference to a past age without coal was made, which was still remembered and, therefore, could not have been too long ago.

It is not easy to formulate a quantitative estimate of actual coal use at that time. Therefore, the following figures only indicate orders of magnitude and it is possible that an increase in the period can be explained in part by improved documentation. This is always the case for quantitative information in periods further back in time, unless one is dealing with a regional case that was by chance more thoroughly documented though then it would not be representative for the entire country. Figures such as those for annual coal production in Great Britain are necessarily extremely imprecise and merely show a trend.

---

*This can only mean that coal was used in blacksmithing. Iron smelting with coal only succeeded in the 18th century.

1551–60	1681–90	1750–60	1781–90	1801–10	1841–50	1895–99	1913
0.2	2.9	4.3	8.0	13.9	46.3	205	275

Table 8. Estimated annual coal production in Great Britain
(Million tons. After Nef 1932, 1, 20; Pollard 1980, 229)

A logarithmic representation of the process of growth permits recognition that this was exponential growth whose rate was only a little smaller in the early 18th century than in the periods before and after. Substitution of coal for wood first began in the trades where large amounts of process heat were required and where no technical problems occurred. Later, the transition to the use of coal was made successively in most commercial applications. Before the 16th century this was only the case in limestone burning, salt making and metal smithing. Here, too, charcoal was preferred but the price difference of these fuels was such that the disadvantages of using coal were worthwhile. Particularly in metal working, there was a risk of losses in quality with the use of coal because its sulphur content discoloured the metal or made it brittle if the heat was too high. Charcoal was used up to the 19th and even the 20th century for particularly fine and demanding alloys.

Large quantities of metal were by then smelted and processed with the aid of coal. However, difficult technical problems prevented the use of coal in iron smelting and an economically useful process on a larger scale was only established in the 18th century. From the middle of the 16th century and especially in the 17th century all kinds of trades experimented with the use of coal instead of wood.

In salt making, where the demand for fuel was extremely high, no particular technical problems occurred with the use of coal because the brine evaporated in open pans. In 1605 approximately 50,000 tons of coal were consumed in salt making in Durham and Northumberland. Approximately 6 tons of coal were required to make 1 ton of salt, depending on the salt content of the brine. Salt consumption rose enormously at that time, because it was used as a preservative, for example of fish, and the demand for salt pork increased with the expansion of seafaring. The consumption of fuel rose likewise and could be covered more readily and less expensively this way.

In most fields where coal was used to generate process heat, the fundamental problem was that it contained chemical elements that could diminish the quality of the processed good when compared to charcoal. Undesirable chemical reactions and the transfer of escaping gases to the product had to be avoided. The solution consisted principally in separating coal from the processed good or by removing its damaging qualities by coking, a process analogous to making charcoal of wood. In those cases production technology often had to be changed entirely, which could cause more or less serious difficulties. There were problems in the following applications: roasting malt for brewing, baking bread, brick making, pottery, metal processing (even apart from iron smelting), glass making and, finally, domestic heating.

The solution usually consisted in developing special firing ovens that were suited to the particular purpose. By the middle of the 17th century these problems were basically solved in the most important applications. Even in brewing and dyeing, where difficulties persisted for some time, the transition to coal became possible in the early 17th century, though varieties particularly low in sulphur were coked.

When efforts were made in Germany in the 18th century to change to coal burning, it was possible to learn from the English experience, though it cannot be overlooked that solutions to particular problems had to be found empirically and so could not always be imitated under different circumstances. Much depended on local conditions, not only in the types of coal used but also in the chemical composition of other raw materials and products. Still, England served as a lesson that these difficulties could be overcome in principle even though modifications were required in any specific case.

With the exception of iron smelting, coal was used by all important trades in England from the 17th century on. According to Nef's calculations, annual consumption in that period was approximately one million tons. Remaining production was exported or used in private households. The dissemination of coal technology played an important role in construction as well. When the process of firing bricks with coal was mastered in the early 17th century, it became possible to build more brick houses, thereby reducing dependence on building timber. After the Great Fire of 1666, London was rebuilt in brick. The production of such a large quantity of bricks with firewood alone would surely have encountered problems.

Resistance to coal burning

As in the Middle Ages, there was a multitude of complaints during the second run at using coal. This was particularly so when it was used for heating rooms or cooking. Resistance lasted well into the 17th century but could not seriously impede the use of coal because of its obvious advantages. In the face of price increases for firewood, the poor were the first to be forced to use cheaper coal. In well-to-do households it was despised for some time and even considered a mark of poverty. This was made evident in a remark of the year 1632:

Within thirty years last, the nice dames of London would not come into any house, or room, where sea-coals were burned, nor willingly eat of the meat that was either sod or roasted with sea-coal fire (Stowe 1632, 1025).

It remained a sign of an elevated life style to burn pleasant-scented wood pieces in an open fireplace even when hearths had been fitted for coal to eliminate most noxious smells. At least the bedrooms were heated with wood by all that could afford it.

It is evident that the widespread use of coal in London was a nuisance for contemporaries, because it was fought in print. Not only sensitive ladies were offended by the smell; pamphleteers, too, took stand against it. John Evelyn is a pre-eminent example. Among other matters he presented a project to promote tree planting, which aimed at countering the scarcity of wood in shipbuilding (Evelyn 1664). Evelyn published a pamphlet in 1661 with the significant title *Fumifugium, or The Inconveniencie of the Aer and Smoak of London dissipated.* This pamphlet employs widespread prejudices in a verbose attack on the polluting stench of coal, articulates them in public and designs a project for relief. It opens with the pathetic complaint:

> That this Glorious and Antient City, which from Wood might be rendered Brick and (like another Rome) from Brick made Stone and Marble; which commands the Proud Ocean to the Indies, and reaches to the farthest Antipodes, should wrap her stately head in Clowds of Smoake and Sulfur, so full of Stink and Darknesse, I deplore with just Indignation (Evelyn 1661, preface).

Evelyn had two objections against coal burning: one medical, the other aesthetic. In the first place he assumed that the use of coal had a direct hazardous effect on health. It was also to be feared that the character of the inhabitants of London might change under the influence of the poisoned air since he, following the Hippocratic tradition, saw a close link between climate and a people's character. As an important indicator of the effects of coal smoke he cited the fact that flowers no longer grew in the gardens and there were no longer bees in the city. All London was polluted and that affected people, who suffered from all sorts of diseases of the air passages:

> Is there under Heaven such Coughing and Snuffing to be heard, as in the London Churches and Assemblies of People, where the Barking and the Spitting is uncessant and most importunate? (ibid., 10).

Not only people suffered from coal stench but all sorts of objects as well. The buildings were blackened, clothing was dirtied and much damage was caused in houses because valuable furniture and luxury goods were destroyed, artworks discoloured and tapestries eaten.

It is this horrid Smoake which obscures our Churches, and makes our Palaces look old, which fouls our Clothes, and corrupts the Waters, so as the very Rain, and refreshing Dews which fall in the several Seasons, precipitate this impure vapour, which, with its black and tenacious quality, spots and contaminates whatsoever is expos'd to it (ibid., 6).

As relief from this pervasive environmental pollution, Evelyn suggests banning trades that use much fuel from London and settling them where either sufficient firewood was available or the gases could not cause such damage. In particular, he was thinking of brewers, dyers, soap and salt makers, lime burners and similar trades. Coppices were to be established in the vicinity of the city to cover its firewood demand.

Of course, these suggestions were not taken seriously. On the contrary, during the 17th century coal consumption increased considerably. Apparently people only gradually became accustomed to the general air pollution and stench. Even as late as the year 1700 Tim Nourse published in an appendix to a text which, as part of the 'discourse on husbandry', concerned itself with all sorts of improvements in agriculture and the domestic economy, a project that proposed to eliminate coal burning in London. His complaints about the nuisance caused by coal are strongly reminiscent of Evelyn's, though his descriptions of effects are more colourful. He considered soot, which he did not distinguish from smoke, as the significant cause of pollution. This 'powder' lay everywhere in the town, filled the air and turned into black, all penetrating slush when it rained. It was considered most unhealthy:

When Men think to take the sweet Air, they Suck into their Lungs this Sulphurous stinking Powder, strong enough to provoke Sneezing in one fall'n into an Apoplexy (Nourse 1700, 350).

Nourse also complained that the most varied objects were destroyed and spoiled, not only fabric and furniture but even iron grids: 'The Stones themselves run the same fate' (ibid.). Only a return to firewood would help.

In a time when the politically and economically interested public expected that such projects contained quantitative information in the sense of 'political arithmetics', Nourse had to prove with figures that coal could be replaced by wood. To estimate the firewood requirements of London, which were not known statistically, he took a detour over population size. Beginning with the (known) annual mortality, he calculated – with an estimated life expectancy of only 20 years – the population figure of London. He obtained 400,000 to 500,000 inhabitants, which may have been reasonably realistic. From this he estimated the number of households, from which he calculated the number of hearths, which he set at 360,000 fireplaces, of which each consumed about one load of wood (= c. 1.2 m³). If all trades were settled outside the city, the remaining annual wood requirement would be roughly 350,000 loads.

The coppiced area which could cover this demand sustainably would have to be 60,000 acres (= 240 km²) according to his calculation. Nourse thought that it should be possible to make such a plantation available, given the disadvantages of coal burning. However, to assure this a number of governmental measures were necessary: forced enclosure of commons to be able to reforest systematically, strict control of woodland use to prevent destruction and exploitation as well as fixing the wood price. As subsidiary effects, he envisioned that the stakeholders in the commons would benefit from the payments and that governmental revenues would eventually rise, since the fenced woodland bore a taxable rent. Here we have the classical argument of a pamphleteer, who will prove that the proposed measure benefits all and harms none.

He even had an answer ready for the last objection – that pollution was merely shifted from the town to the countryside and that making charcoal also caused noxious smells. After all a charcoal kiln could only be established in a single location every twelve to thirteen years,

> ...so that the inconvenience may easily be born with, and is incomparably less than the continual stink of the Sea-Coal Fires, which are so great an Annoyance to the Court, to the Nobility and Gentry, and in a word, to the most Considerable part of the Kingdom, whether we consider their Number or Quality (ibid., 357).

Even this last appeal to the aesthetic and health interests of the upper class could not lead Nourse to success. Obviously coal had become so important in England that its use could not be refused because of such concerns. In the struggle for the dominant position among the European trade nations nobody wanted to give up that decisive trump that was available in the form of coal deposits and their easy transport.

Coal as trump in the trade war

The struggle among European mercantile nations for dominance was the general political background of the historical transition to using coal as an energy source. This struggle for power and wealth was understood as a zero sum game by participants: not only the total sum of power but also wealth in natural resources was considered constant and its amount could not be increased. An increase in wealth in the sense of an increase in productivity was not plausible in the context of agrarian society, since all revenues derived in the end from a fixed land area that could not be increased but only redistributed. Therefore, economically only the distribution of an existing stock was possible, and that applied in particular to trade. From this perspective, the profit of one merchant was identical to the loss of another. Mutual advantage in trade could only arise in respect of the utility of goods but not of their value.

The expansion of power, which can in fact only be understood as a zero sum game, was promoted by the nascent sovereign states by all available means. If the genesis of the modern territorial state must be understood as the successful struggle of a single ruling house against a series of rivals, this was also true on the larger stage, which we may call 'international' from the perspective of the modern nation state. Since the late Middle Ages enlargement of power has been well within the logic of the development towards the modern state.

Everything else was subordinate to this purpose as a mere means. In particular, mercantilist economic policy can be read as a strategy for the expansion of power. Wealth, especially gold, was a tool with which the power of one state could be enlarged relative to its rivals. If a state succeeded in drawing the greater part of the planet's wealth to itself, e.g. in the form of precious metals, it would not only increase its own possessions but, at the same time, reduce those of a potential opponent. International trade was a covert form of warfare: the working population was the army in the trade war and national wealth the arsenal.

In terms of this logic, material production in agriculture and manufacturing could only serve to unfold the means by which a state gained comparative advantage. If domestic trade by definition only served to enhance the wealth of individual merchants, it could not contribute to the wealth of the nation as a whole, since its effects were cancelled out on balance. It could only serve to create the preconditions for foreign trade, which alone was profitable. The widespread theory of profit upon alienation claimed that in domestic trade an individual could only make a profit at the expense of another. In the end, the state need not care about this. However, if individuals profited from foreign trade, this was at the expense of foreigners and the relative profit of one's own country rose. Domestic trade and domestic production were worthy of support only in terms of relative improvement against other countries. For this purpose, they were an important subject of governmental economic policy.

The power struggles of Europe's dynasties, from which the nation states arose, and later the international competition of sovereign states, were tremendous motivation for expanding economic power. The economic development of Europe, which eventually flowed into the Industrial Revolution, is the last but not the least result of these efforts of emerging absolutist states to expand their power. In this context great efforts were made to expand population numbers. The growth of European population, which had taken an upswing since the 17th century, was at least strongly favoured by the demographic policy of the mercantilist state, if not even caused by it to a large extent. A large population was desirable for several reasons: it equated with more recruits for warfare and seafaring, increased the tax basis of the state and helped in attracting a larger share of the world's wealth through increased manufacture. It was not only in Germany, where it was desired to make up for population losses caused by the Thirty Years War, that there was a deliberate policy of *Peuplierung*.

England, too, broadly fits this description but during the critical period it was not an absolutist state in the continental sense. No later than the Glorious Revolution of 1688/89, a system of compromise had developed between the different factions and layers of the ruling classes, who were represented in both houses of Parliament and considered themselves representatives of the nation. Like the leading groups of the United Provinces of the Netherlands, they understood their common interests as the Common Good, and acted accordingly. In their own country, they were dependent to a large degree on entering into permanent compromises, since no group could rely on enforcing its special interests as the presumptive interests of the whole over the long run, as was the case in some countries on the continent.

This political willingness to compromise was economically equivalent, by and large, to letting private self-interest work, so that the market could differentiate into the important economically integrating and socially synthesising agents. The fact that a political structure of checks and balances developed in England, as none ever had before, permitted the market principle to become effective. In terms of world history this meant that the market originated only here as the potentially single centre of economic interaction – and made it possible that a development set in spontaneously and experimentally, to be copied later through political measures by other countries.

In terms of the substitution of wood by coal, this meant that the process was mediated through the market alone in England and was neither seriously sponsored nor inhibited by the state. Projects like those of Evelyn and Nourse not only never stood a chance because of the primacy of the market (which by certainly did not apply everywhere) but were also negated by the generally accepted goal of political economy, of achieving commercial power. A politically enforced rejection of the use of coal would have been equivalent to stringently limiting England's export base.

Land gained with coal use

The foundation of Britain's orientation towards export in the early modern period was the cloth industry. The precondition of this development was created in the 16th century when landowners increasingly went about converting arable into sheep pasture. In the course of this process a layer of landless labourers was formed for whom agriculture offered ever fewer opportunities of employment. This proletariat mostly found new jobs in the textile sector. Since this change of land use was not initially accompanied by a comprehensive improvement of agricultural productivity – which only rose in the late 17th and 18th centuries – England, which had exported grain in the Middle Ages, now had to import grain on a large scale. Thus, the country tapped into international trade flows in a way that permitted it to place a progressive emphasis upon commercial expansion. In this context interesting shifts in land use came about.

Land area was the standard measure of agricultural societies. Energy was collected on the surface of land, land was the source of the landlords' revenues, its ownership conferred political power and wealth was always wealth in land. The surest way to increase wealth consisted in increasing the area in possession. This could occur through direct taking of land, as in the case of the *Reconquista* or *Conquista* of the Spaniards, where in the ideal case not only was land conquered but also a labour force was subjected to cultivate it. Another form of gaining land lay in making available what can be called 'virtual area' (cf. Borgstrom 1972). This means the provision of goods for the production of which a certain share of the territory would have to be used that could be dispensed with if these goods were imported.

One example of the gain of virtual area is that of fisheries: Europe is a peninsula with a long coastline that favours fisheries. Whaling started in the Middle Ages with the Basques in the Bay of Biscay, and expanded from there to Greenland and Newfoundland. Not only human food was caught in the form of whale meat, but also oil as fuel for lamps, softeners for textiles and leather and machine grease. In the early modern period the herring fishery gained in importance, especially in the Netherlands. When fish was caught in the oceans, the protein supply of the fishing population was improved. The same effect could only be achieved domestically if cattle were kept for slaughter, but pasture would have to be made available. We could attribute a certain acreage to a quantity of fish, which was no longer needed for meat production but could be used for other purposes. This meant that the total available land area of a fishing nation (such as the Netherlands or Great Britain) could regularly be increased by this fictive amount without counting the seas formally as land area.

Of particular interest are commercial areas. These are virtual areas gained by importing goods which would have required larger areas than were needed for the production of exported goods which financed these imports through sales. In the first instance the boreal forests of northeastern Europe may be cited as such commercial areas for Britain and the Netherlands. Russia had exported furs, honey, tallow, blubber oil (from seals), sturgeon, flax, hemp, pitch and salt to the west from the 16th century. The area of overseas colonies must be added, which primarily produced luxury goods such as sugar cane, tobacco, cotton, indigo and above all spices.

Jones (1987, 82, following Webb 1952, 18) estimated that the area available to Europeans due to colonial expansion in the 16th century theoretically increased from 24 acres per head to 148 acres per head. In particular the conquest of the Americas led to a dramatic increase of territory, as becomes very clear in a comparison of European colonial powers with other agrarian civilisations of Eurasia.

If Russia west of the Urals is counted as part of Europe, the total landmass is about 10 million km². If a European population of 80–100 million is assumed around 1500, this would be 8–10 ha per person. Apparently, these are the figures Jones is using. However, in 1500 no one would have thought of counting Russia as part of Europe. Europe without Russia has an area of only about 5 million km², while its population was probably 70 million around 1500 (Mols 1974; Livi-Bacci 1997, 31). Theoretically that would have been about 7 ha per person in Europe or a population density of about 14 persons per km².

It is far more difficult to estimate population outside Europe. In the colonies the indigenous populations must be included. The American continent has a total area of around 42 million km². If more recent estimates are right that about 80 million people lived in the Americas around 1500 (the estimates fluctuate between 10 and 100 million, cf. Stannard 1992; Kiple 1997), the combined population density of Europe and (uncolonised) America would have been about 3.1 persons per km². Around 1800 the population density would have risen to 3.6 persons per km² but with a decisive shift in proportion: 146 million Europeans against only 24 million Americans (Livi-Bacci 1997, 31), who were now to a large extent the descendants of migrants (immigrants from Europe and slaves imported from Africa). According to our calculation the population density for Europeans would have diminished to a third (from 14 persons per km² to 3.6 persons per km²) if we consider America as part of their living space.

Year	India	China	Anatolia	Europe	America	Europe + America
1500	23	25	8	14	1.9	3.1
1800	42	80	12	29.2	0.6	3.6

Table 9. Population per km², 1500 and 1800

In principle, Jones's conclusion is plausible, that in the areas controlled by Europe the population density declined, despite considerable demographic growth during the early modern period if colonial territories are factored in. Important in this are the mass deaths of indigenous populations in the colonies. Furthermore, it should be kept in mind that the territories remained largely unused and their possession initially hardly had any influence on the lives of the masses in Europe. Wallerstein's thesis (1974, 44) that the population of Europe was already metabolically dependent in the early modern period on colonial imports is not credible. Until the 18th century colonial imports were almost exclusively luxury goods (sugar, tobacco, rum, spices) of no significance to the dietary budget of the masses. Grain imports from America, South Africa, Australia and Ukraine only achieved significance in the 19th century.

The only colonial product that could have played a metabolic role in the nutrition of the European populace was sugar. The Moors had cultivated sugar cane in Andalusia in the Middle Ages and there were sugar cane plantations on which African slaves worked in Cyprus from the 13th century. This pattern was extended to Sicily, then to Madeira, the Canaries and the Cape Verdes by the Spaniards and Portuguese and finally to the West Indies and Brazil.

Sugar cane was first cultivated in the Old World and reached America in the course of European expansion. However, in terms of quantity the consumption of sugar remained a marginal matter for some time. The sugar isle of Cyprus only produced 100 tons a year towards the end of the 15th century. The great age of sugar production began after 1680, when Caribbean islands such as Martinique, Guadeloupe, Curaçao, Jamaica and Santo Domingo were used. Here the classical system of sugar plantations based on slave labour was established. Only then did true mass production come about. In Santo Domingo it amounted to about 70,000 tons annually in the 18th century.

Based on documented production figures, the nutritional importance of sugar to the population of Europe can be roughly estimated. Great Britain annually consumed about 10,000 tons of sugar around 1700, but about 150,000 tons around 1800 (Braudel 1974, 157). The British population (including Ireland) grew in this time period from approximately 9.3 to 16 million. Therefore, the per capita consumption of sugar rose in the course of the 18th century from 1 kg to 10 kg. The caloric value of 1 kg sugar is 16 MJ, which is equivalent to human energy requirements for 1.5 days. This means that the role of sugar in the nutrition of the British population can still be ignored in the early 18th century. However, around 1800 sugar imports contributed 4% of the calorific needs of the British population or, in other words, the grain acreage of the British Isles would have had to be increased by 4% to produce the same quantity of food calories domestically.

In pre-revolutionary Paris annual sugar consumption was around 5 kg per person. In all of France apparently no more than 1 kg per person per year was consumed in the late 18th century. Sugar was expensive but ubiquitous even in peasant homes, where a sugar cone hung on a string from the ceiling over the table and beverages were quickly dipped at will. However, it was still a luxury good and not a foodstuff of significance to the dietary budget. Wallerstein's opinion (1974, 43), 'one of the most important complements in the European diet is sugar, useful both as a calorie source and as a substitute for fat', only applies to the 19th century, when large amounts of colonial sugar cane were imported and sugar was made from beets in Europe. But this was already within the context of the fossil energy regime.

It is apparent from these figures that real gains in territory were much smaller than a look at the map suggests. The colonial territory existed but could not be used extensively in the context of an agrarian society. America was as good as empty after the collapse of indigenous populations in the 16th century, but without providing any noticeable relief to European populations. European migration to America in the entire period before 1800 only comprised 2 million people. America did not become a substitute space for European industrial countries until the 19th century during the Industrial Revolution. Only then did migration on a scale worth mentioning occur. The fertile grain producing regions of the Mid-west were opened and only then were larger quantities of food exported to Europe: grain from the USA, meat from Argentina.

The background to this was the transportation problem. Potential grain growing areas in the North American Mid-west were as inaccessible to consumption centres as the Argentine pampas before the railway. The prohibitive expense of land transport has been mentioned before. But even the advantages of water transport were not unambiguous when large quantities of goods and large distances were involved. Without a doubt there were tremendous innovations – in stability, navigation, steering and armament – in seafaring on the high seas during the early modern period, which permitted regular commercial traffic on the oceans; but the size of ships did not change significantly. Already in the 16th century Genoese ships had a capacity of up to 1,500 tons and Venetian vessels of 1,000 tons transported cotton from Spain to the Near East. Portuguese ocean-going ships of this age had a tonnage of up to 2,000 and this was still the upper limit of British East Indiamen in the 18th century. Even in the early 19th century many ships merely had 30–50 tons capacity and, as a rule, freight ships in overseas traffic had 200 tons. These dimensions only changed with the construction of steel ships and the victory of steam ships fuelled by coal. Only then did large areas of America and Australia really become accessible to Europeans.

Let us return to early modern England. Its commercial expansion had both a technological and a territorial aspect. Arable was turned into pasture and wool was processed into cloth and exported. Grain, which could not be domestically produced in the same quantity since less arable was available, could be imported with the export revenues. International exchange only made sense if more grain could be traded for the cloth than could have been grown on the pasture on which the wool was produced, or if the textile sector used less labour than agricultural production. A precondition was that the productivity of the commercial sector of England was higher than that of agriculture. At the same time it had to be more favourable for the grain exporting countries to trade grain for English cloth than to produce it themselves. With this international division of labour there had to be different natural conditions or a technological gradient that enabled the commercial country to extend its de facto territory across its boundaries. In England the latter was case. In a sense, land was imported as grain and knowledge was exported as cloth, thus increasing the ecological carrying capacity of the British Isles: with the help of foreign grain a larger population was supported than would have been possible restricted to the domestic territory under the given agricultural technology.

This process, of course, is mediated socially and therefore somewhat complicated. It had to be more profitable for a landowner in England to keep sheep than to grow grain and that independently of the fact that wool could be turned back into grain through trade. If the interest of the proprietors alone were at stake, the sheep might as well have eaten people, as Thomas More feared. A reason for the high profitability of sheep raising may have been that they were associated with lower labour costs than cultivation. That the landless poor earned their wages and bread through commercial expansion was a coincidence that did not concern landlords. It could have happened – and it did and still does in occasional instances – that wool or cloth were only transformed to a lesser degree into grain and only to the extent necessary to feed highly productive skilled workers, while the bulk of exports is used to import commercial raw materials and luxury goods for the upper class. Such a development is fatal for the landless and unemployed poor. At the beginning of the historical process of industrialisation this did not happen, because productivity was relatively low, production was labour intensive and only a small amount of raw materials relative to their overall use was imported.

The following link must be emphasised: the shift in land use structure and the associated technological and commercial developments in England were favoured by converting into pasture that portion of area formerly used to furnish firewood, since coal could substitute for wood. The arable did not have to be reduced to the full extent that would have been necessary without coal; or, expressed somewhat differently, more pasture was obtained this way. Fossil energy freed land, which greatly favoured commercial development.

To make apparent the importance of the new energy source, coal, to England's position in the European world market, the material flows between Northeastern and Western Europe will be considered first. It should be noted for a better understanding of the table that western Europe is not only Great Britain but includes the Netherlands, while other areas did not play a large role in these commercial relationships.

Eastbound	1621	1721
Salt	24%	14%
Fish	21%	10%
Wine	8%	13%
Colonial Goods	24%	31%
Textiles	21%	28%
Other	2%	4%
Westbound	**1621**	**1721**
Grain	59%	28%
Forest Products	10%	27%
Hemp and Flax	14%	13%
Metals and Ores	15%	29%
Other	2%	3%

Table 10. Material flows in the East/West trade 1621 and 1721

(Bogucka 1978, 41; cf Aström 1978)

In our context two developments are of particular interest. Firstly, the import of forest products to western Europe increased, in particular of wood for beams, masts, planks and other boards, and also of pitch, tar and resin as well as shipbuilding materials. On the other hand, luxury goods such as colonial products and wine were increasingly transported to eastern Europe. From the material point of view, it can be said that England and Holland built ships with materials from eastern Europe (which included hemp and flax for ropes and sails) to conduct colonial trade that made it possible to pay for those materials and to feed the population employed in this undertaking. The flow of these materials through England and Holland enabled these countries to develop a structure that probably would not have been viable based on native resources alone.

This becomes particularly clear when the import and export of grain is considered. The share of grain in exports from eastern Europe to the west halved within a century. England even became a grain exporter again in this time period. In 1640 the share of agricultural products constituted only 1.4% of English total exports. From this time grain export from England grew enormously: from 2,000 quarters in the decade after 1660 it rose over 300,000 quarters in 1675/77 to 900,000 quarters in 1745–54 (Coleman 1977, 65). This happened despite population growth in the British Isles from 6.8 to 9.3 million between 1600 and 1700. If one considers that at the same time a large part of the English land area was used as pasture, it is apparent that such strong pressure was exerted on existing forests that their share in total area diminished considerably.

In grain cultivation an increase in productivity undoubtedly formed the basis for the expansion of export, but elimination of the competing land use for coppicing surely also played a role. What this meant is easily recognised if we consider real land use in England and Wales during the Industrial Revolution. For this purpose we would like to use the results of a summary based on contemporary estimates:

	1700	1800	1850
arable	29	30.1	39.1
pasture, meadows	26.3	45.4	42.8
woods, coppices	7.9	4	4
forests, parks, commons, waste	34.2	16.8	8
buildings, water, roads	2.6	3.3	5.8
total	**100**	**99.6**	**99.7**

Table 11. Land Use in England and Wales

(% of total area. Source: Allen 1994, 104)

For our purposes this table requires an interpretation. The three basic types of agrarian habitat (arable, pasture, wood) are clearly differentiated, though a residual category is added here in which wood and pasture land are lumped together. Commons is a legal category, waste is an agronomical category, parks is an aesthetic category and forests are difficult to distinguish from woods and coppices. Despite these conceptual difficulties the following historical trend is recognisable: around 1700 coppices made up 7.9% of the area but by 1800 their share had almost halved. Around 1700 forests, parks, commons and waste made up a third of the area, by 1800 they still were 16.8% and by 1850 barely 8%. Since it may be assumed that fuel was also obtained from these areas, the overall share of woodlands dropped even more than is obvious from the decline of woods and coppices.

If, by contrast, we assume that arable, pastures/meadows and woods/coppices are intensively used agricultural areas, then it is clear that the colonising use of the overall area rose dramatically: 63% of the overall territory was used for agriculture by 1700, by 1800 nearly 80% and about 1850 over 87%. At the same time the share of firewood land fell drastically: from 12.5% around 1700 to just over 5.2% around 1800 and only 4.6% around 1800. This can only be interpreted in the sense that wooded areas were freed by the use of fossil fuels and so could be used for another purpose. If there had been no coal, the demand for firewood would have risen dramatically at the expense of other land uses. Following the classical locational considerations of Thünen, one might even expect that the coppiced area would have increased more strongly while grain cultivation would have retreated due to the fact that the relative share of transportation costs was lower for grain than for firewood. With a shortage of firewood there would have been more reforestation, and grain, instead of wood, would have been transported from overseas.

An essential precondition for the decline of coppices was the introduction of coal burning. It can be said that territory became available for other forms of land use because of the substitution of wood by coal, which permitted an elevated standard of productivity and provisioning. Territory for energy plantations – coppices may be so defined – was not required to the same extent as before since the pertinent demand for energy was covered by coal. An attempt will now be made to estimate the size of the virtual area gained by the use of coal, i.e. a certain quantity of coal will be equated with an area that would have been required for the production of firewood in its place.

To illustrate how much land was freed for arable and pasture by using coal, I have used the figures from a German inquiry of the 18th century. It reports experiments that were undertaken to compare the combustion values of wood and coal. These figures appear suited for a historical comparison because they reflect the efficiency of fuel use in the period. A comparison with actual physical values demonstrates that they were by no means unrealistic.

The investigation was carried out by Samuel Hahnemann, the founder of homeopathy, and differentiates two types of wood. According to it one cubic foot of coal has a combustion value that is equivalent to the following amounts of wood:

pine	7 $^6/_7$
willow	8 $^1/_2$
poplar	13 $^2/_5$
birch	7 $^1/_6$
alder	7 $^2/_5$
European beech	6 $^1/_4$
European hornbeam	7 $^1/_7$
oak	6 $^1/_2$
aspen	11 $^1/_2$

Table 12. Volumes of wood equivalent to 1 cu. ft. coal
(Cubic feet. Hahnemann 1787, 6)

On average, one volumetric unit of coal is equivalent to 7.8 volumetric units of wood. If we assume a specific weight of 1.5 g/cm³ for coal (which is problematic since they were 'cubic feet' of wood and coal that lay in a basket and were not packed tightly), then 1.5 tons coal = 7.8 m³ wood, that is 1 ton coal = 5.2 m³ wood. A similar result is obtained with actual physical figures (Smil 1991, 323): wood has a specific weight of 0.4–0.7 g/cm³, so 1 ton wood equates with approximately 2 m³. The combustion value of anthracite is about 30 MJ/kg, that of firewood about 12 MJ/kg. This means that 1 ton coal has the same combustion value as 2.5 tons or 5 m³ wood. We will subsequently use these figures as base values, while remaining aware of the fact that various types of wood and coal have different combustion values that may deviate from each other by as much as 50%. Most importantly, these physical values say nothing about the energetic efficiency of combustion. It can be as low for coal as for wood, that is, both fuels may have tremendous potential to increase technical efficiency.

As we have seen, 1 ha of coppice has an annual yield of 5 m³ wood. Under stationary conditions, when only the sustainable yield of a forest is considered, an annual use of 1 ton coal makes 1 ha land available that would otherwise have been required as fuel plantation. If we transfer this to known figures of historical coal consumption, the ecological significance of the transition to coal becomes readily apparent (Table 13).

We may now compare these figures to land mass. The entire area of England, Wales and Scotland is approximately 228 km². Already in the 1820s, British coal production freed an area that was equivalent to the total surface of Britain! This is even more dramatic if we consider England and Wales alone. The area of Scotland comprises a third of the whole, while Scottish coal production

Annual average	Coal million tons	Wood million m³	Area 1000 km²
1551/60	0.2	1.0	2
1681/90	2.9	14.5	29
1751/60	4.3	21.5	43
1761/70	5.2	26	52
1771/80	6.4	32	64
1781/90	8	40	80
1791/1800	10	50	100
1801/1810	13.9	69.5	139
1811/1820	17.5	87.5	175
1821/30	22.6	113	226
1831/40	32.3	161.5	323
1841/50	46.3	231.5	463
1855	64.5	322.5	645
1860	80	400	800
1870	110.4	552	1104
1880	147	735	1470
1890	181.6	908	1816
1900	225.2	1126	2252
1913	287.4	1437	2874

Table 13. British coal production and its possible substitution by wood
(Figures on coal production: Buxton 1978, 86; Pollard 1980, 229)

only comprised 15–20% of the British total. Coal production in England and Wales already amounted to about 15 million tons around 1810, which was hypothetically equivalent to the area of England and Wales (150,000 km²). This meant that the energy capacity of the agrarian solar energy system in the heartland of industrialisation was already exceeded in the first years of the 19th century in the thermal energy sector alone (Wrigley 1988, 52 arrives at similar conclusions). The availability of fossil energy catapulted Great Britain into a novel economic energy regime.

What this meant can be illustrated by the following diagram where the virtual area won by coal consumption is added to the different forms of land use as they developed between 1700 and 1850. It becomes clear why the areas which provided for wood as a fuel could decrease so dramatically during the Industrial Revolution – they were replaced by the virtual area of fossil fuel.

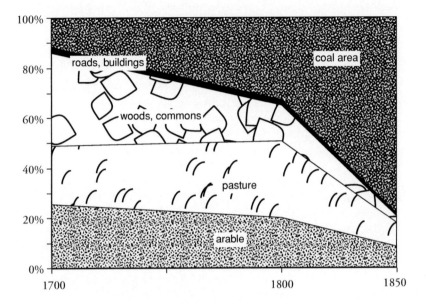

Figure 3. Structure of land use in England and Wales

Was England dependent on coal?

Let us consider what it would have meant if Great Britain had not possessed coal reserves because of an accident of geology. Today the example of Japan demonstrates that successful industrialisation is possible without a domestic energy base. Is it conceivable that the missing amount of energy could have been imported in the form of firewood or of goods that were produced with a relatively large amount of wood? Were such quantities of wood actually available and where would they have come from?

First, it should be noted that Britain did import large quantities of wood and wood-dependent products: timber, ship masts, pitch, tar, resin, iron, potash etc. In the 18th century, annual British wood imports (masts, beams and boards) fluctuated between 1.5 million and 8 million cubic feet, i. e. 40,000 to 200,000 m³ (Aström 1970, 20). There are no exact figures on the import of wood-dependent products but if one assumes that 20,000 tons of iron were imported annually in the early 18th century, that is the equivalent of one million m^3 wood in the year. Roughly estimated, the actual import of wood and wood products should have been at least 1.5 million m^3 annually, of which the smaller part was shipped as wood and most as products that consumed great amounts of wood in their manufacture. The wood imported into England was primarily employed in construction, especially in shipyards. Firewood was not imported except used as ballast.

English imports of wood and wood-dependent products came primarily from Scandinavia and northeastern Europe (Russia, the Baltic, Poland). Towards the end of the 18th century there was a shift from Norway, where the first signs of an exhaustion of wood supplies apparently appeared, to the Gulf of Finland region, to St. Petersburg (Neva), Riga (Dvina), Memel, and Danzig (Vistula).

Between 1760 and 1810 raw material prices for wood in northeastern Europe rose faster than freight rates. In Finland they rose by double the rate of other prices (Aström 1978), which may be taken as a sign that easily felled stock was declining. I do not have exact figures for Finnish forestry in the 18th century but since the region around the Gulf of Bothnia were still fairly unexploited, information from the 19th century may be useful (Lagus 1908). Towards the end of the 19th century Finland possessed about 15 million ha of forest, with the annual yield lying between 0.8 and 2 m^3/ha. This is clearly less than the 4–8 m^3/ha with which one reckons in central Europe and is due to less favourable soil and climate. 1.55 m^3/ha is given as the average annual yield. The amount of wood that could be sustainably used lay theoretically around 23 million m^3 annually, but given difficult access and a low level of forestry technology in the 18th century, less than 20 million m^3 annually may be assumed.

When lumber is prepared for export about 40% waste is produced in the sawmills. This may be used for pulp and paper production and fuel nowadays, but under the technological conditions of the 18th century this was not yet possible. On the other hand, it was advisable to export lumber because it was easier to transport boards and squared beams, which required a smaller loading volume, and because of the greater water capacity of sawmills in Scandinavia.

If we assume an annual increase of 1.55 m^3/ha and an average forest age of 100 years, a total wood reserve of approximately 2,250 million m^3 results. It should be noted that spruce trees can reach an age of around 300 years but hardly grow in the later years. Roughly estimated, the forest untouched by humans was about one hundred fold the annual growth. We shall assume in our calculations that only the annual yield was used and momentarily ignore the energy capital of primeval forests that could only be used once.

How much wood yield could be made available for export? Of the estimated Finnish wood production of 19 million m³ in the year 1900, no less than 13.1 million were used by private households, 0.5 million by the transport system (railways and steam ships), 3.8 million by industry, i. e. 17.5 million m³ domestically. Only 1.5 million were exported. However, the entire yield was not consumed with 19 million m³/year. On the other hand the entire theoretical increase was probably not available because of the well-known problems of transportation. Lagus estimates that the Finnish wood consumption of 19 million m³ actually lay above the annual increase since certain areas were excessively exploited while others largely remained untouched because they were inaccessible. Using an optimistic calculation no more than 5 million m³ could have been exported from Finland in a year.

Let us consider the figures in other heavily forested areas of Europe (after Brockhaus 1875 ff.): Sweden possessed 16.5 million ha wood, Norway 7 million ha, European Russia including the Baltic and Poland 125 million ha, of which much was inaccessible. If we consider an average annual increase of c. 2.5 m³/ha (which is possible under the climatic conditions of Poland and South Sweden), we obtain an amount of more than 400 million m³ wood regrowing annually. According to our calculation this could have substituted for c. 80 million tons of coal, which was equivalent to the British production of 1860. It is tenfold the British annual demand of 1781–90.

However, it was demonstrated in the case of Finland that not the entire yield could be exported, but only a quarter of it. If we transfer this figure to all of northeastern Europe, approximately 100 million m³ wood annually would have been available for export, a quantity that would have replaced British coal production around 1820. If we look at the forest stock of all of Northeastern Europe, the enormous figure of $4–8 \times 10^{10}$ m³ wood results, a quantity that appears truly inexhaustible (it is equivalent to the quantity of fossil fuels used annually and worldwide today).

We must ask if it would have been possible to transport the required quantities of wood from Northeastern Europe to Britain. In the 18th century 1 nautical ton storage room was required for the transport of 1–1.25 loads of wood (= 1.2–1.7 m³) (Albion 1926; Kent 1955/56). The average size of ships that transported wood was about 300–350 tons, with the average ship making four to five trips annually to Norway but managing only two trips to the Baltic (Davis 1962). Therefore, the annual capacity of a ship was 1,800 m³ wood over three trips. For wood transport alone the following number of ships would have to be used:

Year	Ships	Total tonnage
1551/60	555	194,250
1681/90	8,333	2,916,550
1750/60	12,388	4,335,800
1781/90	22,777	7,971,950
1801/10	40,000	14,000,000

Table 14. Hypothetical number of ships in wood transportation

(350 tons each)

We may compare these figures with the actual British tonnage of 1790: the royal navy possessed 300 ships with a total tonnage of 391,450, an average 1,300 t, as was then customary for warships. The merchant navy comprised fewer than 15,000 ships, but with an average tonnage of only 97. Not more than 201 private ships were larger than 400 tons, all of them East Indiamen. In the navy there were ships of more than 2,000 tons. The total tonnage of the merchant navy was about 1.5 million tons. It could have replaced a fifth of coal consumption in 1790 with wood imports if all other imports had been rejected.

The question now is whether it would have been possible to build larger ships to transport these huge quantities of wood. Significant technical reasons contradict this: for the construction of wooden ships specially shaped structures were required for certain particularly stressed parts that had to be made of one piece. The larger the ship, the more difficult it became to find these pieces. The oaks from which they were cut could not grow in a high forest but had to grow solitarily in a park or on a street since only then would the desired forks and bends develop. It was reported that it sometimes took months, even years to find the right tree when large warships were built. Therefore, the Admiralty attempted to limit the size of East Indiamen in order to eliminate competitors. Apparently it was not possible to increase significantly the number of wooden ships of any considerable tonnage. On the other hand, there were 1,400 ships that transported coal at the beginning of the 18th century. In 1695 about a third of British tonnage was used to transport coal. These coal ships were relatively large, mostly over 300 tons (Smith 1961; Dyos and Aldcroft 1974). Certainly they could have been used to transport wood but the stock would not have been nearly sufficient. Towards the end of the 17th century triple their tonnage would already have been required.

Another problem would have been wood use for the construction and repair of the ships. In shipbuilding one load of wood was required for every ton of water displacement while a ship could carry just that amount of wood per ton (Albion 1926, 149). The first shipment of wood would have been used entirely for building a ship. Furthermore, about half the wooden parts of the hulk had to be exchanged annually. If we ignore the original construction and the fact that the rotten wood was perhaps in part useful as firewood and assume that a ship annually consumed a half ton of wood per ton of loading space, it could only deliver one and a half times its tonnage net over two trips. The yield factor for wood transport was about 75%, i.e. a quarter of the wood supplied was consumed in transportation. Furthermore, an additional 20,000 large ships would have surely required a large increase of port facilities, a multiplication of crews and a tremendous supply of sails, pitch, tar, roping etc. There are no figures that permit an estimate.

Finally, at the end of this hypothetical calculation, attention may be drawn to another problem. In the transport of wood and also coal, the value of the load was far less than that of colonial and manufactured goods. For wood the ratio was one to seventy (Albion 1926). Wood or coal was often only taken along as ballast, so that the return journey was not entirely unpaid. Special wood transport was often only worthwhile for wood that was rare or expensive for particular reasons (masts, tropical woods for cabinet making). The basic raw material price of plain timber (beams, boards) was only 5% of the end price, that is how high transport and labour costs were. One can imagine how much the price of wood would have had to rise relative to other products to make it worthwhile to ship wood from northeastern Europe to England just to burn it there.

The preliminary finding of our considerations is that the wood yield of northeastern Europe was quite large, but that it would hardly have been possible to supply England with imported firewood on a grand scale. Would it not have been preferable to import wood-dependent products, which required the input of large quantities of energy in their manufacture? Let us consider the weight of certain goods relative to the weight of the wood that was used in their production:

Salt	1 : 7
Pig iron	1 : 15
Wrought iron	1 : 30
Silver	1 : 200
Glass	1 : 2,400

Table 15. Weight ratios between products and wood consumption

(Gleitsmann 1982)

High figures also apply to other metals and chemical products. In this light it is difficult to imagine that it would have been economical to transport wood from northeastern Europe to England to make salt or smelt iron there. Salt, iron, glass, alum, vitriol etc. would rather have been produced on site and then transported to England. As we shall see that is exactly what happened in the late 17th and early 18th century with wrought iron, the last important product that could not be produced with coal. Other trades, too, would be more likely to develop where an inexpensive energy source was available. Much speaks for the notion that without coal a more decentralised development of trades would have occurred, equivalent to the systemic conditions of agricultural production. A look at iron smelting demonstrates that, apart from this decentralising tendency, industrial production was moving towards a fixed upper limit.

2. Wood and Coal in Iron Smelting

Technical problems

By 1700 all English trades used coal instead of firewood, with one important exception: iron production. There, coal use faced significant technical difficulties that were only overcome after a very long and complicated experimental process. In nature iron occurs as iron ore in which iron oxides (Fe_2O_3; Fe_3O_4 etc.) are embedded in ore-bearing rock. It is essential to reduce the iron oxides during smelting, in other words to break the oxygen bond by applying energy, and to diminish the content of other elements, such as sulphur and phosphorus, as well as the high carbon content. In traditional iron smelting using a blast furnace three phases of processing are differentiated:

i.   Smelting. Iron has a lower melting point than the ore-bearing rock and can be separated from the latter when it liquefies. Most importantly, the iron/oxygen bond is broken in a thermal reaction. The released oxygen combines with carbon monoxide that is flowing through and escapes as carbon dioxide into the air. The result of this process is pig iron, which has a relatively high carbon content and is therefore relatively brittle and breakable. The most important elements of the process are a temperature above 800° C and carbon that burns incompletely to carbon monoxide and, in turn, serves as the base for the reduction of the iron oxides.

ii. Refining. In this step in the process, the iron impurities must be removed, especially the excess content of carbon and other elements that are still bonded to the iron. This necessitates above all a high temperature and a strong, oxygen-supplying draught to permit gases to escape quickly. The iron must be prevented from forming new bonds, so only carbon that is as pure as possible can be considered as fuel. The result is wrought iron or steel. The latter has a carbon content of about 2%, wrought iron of less than 1%.

iii. Processing (Forging). The refined iron is heated and made malleable. The temperatures must not be too high so that no danger arises of the wrought iron unintentionally bonding to other elements and deteriorating in quality.

Apart from carbon, mineral coal contains a number of chemical elements such as sulphur and phosphorus that render iron refining impossible if the unmodified coal comes into contact with pig iron. In principle, smelting is possible with coal, but the danger exists that temperatures rise to the point of causing the silicates in the ore to liquefy and form difficult-to-dissolve bonds with iron. Forging the iron is the most feasible of the processes, but the temperatures must not be too high. Coal was used for forging from an early point in time, even though the product was not always of the best quality.

For the other processes two routes were pursued to make coal useful: attempts were made to purge it of undesirable qualities by coking and efforts were made to separate the coke from the ore or pig iron. Given the very different chemical properties of various types of coal and iron ores, solving technical problems was very difficult. Unlike other industrial projects, iron smelting was not able to draw on theoretical scientific knowledge until a late point in time. Metallurgical processes that pragmatically and empirically developed in the 18th century could only be explained in scientific terms retrospectively, during the 19th century.

From the 16th century attempts were made in all trades, including iron smelting, to replace wood with coal. In the 16th and 17th centuries there were 25 patents for iron smelting processes with mineral coal: 1589 for Thomas Proctor and William Peterson, 1595 for Sir Robert Cecil, 1607 for Robert Chantiell, 1611 for Simon Sturtevant, 1665 for Dud Dudley, 1666 for Hugh Grundy and John Copley. The large number of patents shows that an urgent need was felt for such a process. It is also an indicator that patented processes were not very effective. A successful and industrially useful process was probably first developed by Abraham Darby around 1710 (Flinn 1959a). Even then, it took almost 50 years to establish itself.

Before we consider this question more closely, we shall look at the location and natural facilities of iron works. The smelting process using a blast furnace was taken over from the Continent in Sussex towards the end of the 15th century. An iron works made heavy demands on its location: it had to be close to a deposit of ore to avoid costly transportation. It had to be near running water since it required water power to operate the bellows and hammer mills and because the products had to be transported by water to consumers. Therefore, the iron manufacturing industry was scattered over the countryside and as a rule did not operate near population centres, where wood provision would have been less convenient. In the 16th and 17th centuries the most important centres were located in the Weald of Sussex and the Forest of Dean, in the early 18th century they migrated to more remote areas such as the Lake District, Herefordshire, South and North Wales and even the Scottish Highlands. An essential locational requirement was a supply of wood since iron could only be made with charcoal.

Fuel shortage and stagnation

We demonstrated above that the theoretical upper limit set by natural conditions for iron production in a single location lay around 2,000 tons annually. The figures actually attained were considerably lower. In the Weald of Sussex a single blast furnace produced 200 tons annually from the 16th to the 18th century, in other areas it was 350–450 tons, the actual upper limit lay around 800 tons. There was obviously still room for technical innovations even though the use of charcoal set a maximum for other reasons. It was already known that charcoal must not crumble in a blast furnace, so that the correct air draught would be provided. This limited the possible total charge of a blast furnace. Two blast furnaces could have been built at a single location but that resulted in problems of supplying mechanical energy. Nevertheless, production per blast furnace rose slightly between 1580 and 1720, while the total number remained the same. Overall, British iron production stagnated in this period.

Decade after	1620	1630	1650	1680	1690	1720	1750
Thousand tons	19	20	23–24	20–21	23	25	30–33

Table 16. British iron production

(Hammersley 1973, 602)

Around 1680 iron imports amounted to 16,000–18,000 tons and around 1700 to 16,000–19,000 tons, i. e. they were on the same level as domestic production. English wood scarcity, or more precisely the preference for alternative land uses, led to large scale imports not only of timber but also of iron. It should be considered that the production of a ton of wrought iron required about 50 m³ wood, which was the equivalent of 10 ha land. It can be said that in terms of energy the ecosystem of English industry extended to northeastern Europe from where not only wood but also raw materials requiring large quantities of wood for their production were imported. How long and how far this system could have been extended is difficult to say. Fundamentally, these limits also applied to Sweden and Finland, though in Sweden at least there would have been more rivers to power bellows, iron hammers and sawmills. The situation in Denmark was similar. Though there was ore in the country, iron smelting was abandoned there in the early 17th century due to a lack of wood. Even melting down scrap had to be abandoned. After that iron was introduced from Norway – which belonged to Denmark politically (Kjaergaard 1994, 124).

The question whether a fuel shortage in the sense of scarce firewood was a cause for the stagnation of the British iron industry from the middle of the 17th century has been debated for some time in the literature (Flinn 1958; Hammersely 1973; Clow 1956; Hyde 1973; Hyde 1977). Many misunderstandings surrounding 'wood scarcity' can be reduced to the fact that no clear distinction was made as to whether wood was a free good or an agricultural product. Undoubtedly, there were no woods in England after the 17th century in which wood grew naturally and was free, i.e. was not managed for forestry and did not bear a rent. This was different in Sweden and Russia where iron production possessed a cost advantage. On the other hand, it cannot be disputed that there were coppices in England that were intensively and presumably sustainably used by iron works. Hammersley (1973, 597) notes that 3 out of 10 of all blast furnaces were more than 100 years in operation, and more than half longer than 50 years. This is only possible with a sustainable coppice economy and not by extensive exploitation of stocks. Because these iron works were close to ore deposits and far from population centres and therefore under no strong pressure from alternative land use, forms of forest usage that were stable in the long term could develop. There are contemporary indications:

> It is now in all these parts [Gloucestershire, Midlands] every day's experience, that Gentlemen and others do make their business to inclose Land and sow them with Acorns, Nuts, and Ash Kayes, to rear Coppices Woods, they knowing by experience that the Copice Woods are ready money with the Iron Masters at all times (Yarranton 1677, 51).

Frequently blast furnace owners and forest owners were one and the same. Their economic calculation did not always include rent as part of production costs. Often they only considered wages for forestry workers and charcoal-burners together with transport costs as fuel costs. Wood was considered a free good. This only changed during expansion, when additional land had to be rented or bought. Therefore, it would be absurd to assume a fundamental conflict between woods and iron smelters in the sense that the expansion of the iron industry destroyed forests. Andrew Yarranton clearly wrote in 1677 that the opposite held true:

The next thing is, Iron-works destroy the Woods and Timber. I affirm the contrary; and that Iron-works are so far from the destroying of Woods and Timber, that they are the occasion of the increase thereof. For in all parts where Iron-works are, there generally are great quantities of Pit Coals very cheap, and in these places there are great quantities of Copices or Woods which supply the Iron-works: And if the Iron-works were not in being, these Copices would have been stocked up, and turned into Pasture and Tillage, as is now daily done in Sussex and Surry, where the Iron-works, or most of them, are laid down (ibid., 60).

A lack of wood only existed if it was more profitable to use the land for other purposes than planting coppices. If the iron works destroyed woods, it could only have been by transforming high forest into an energy plantation with short cycles of cultivation. Firewood production for iron works operated at the expense of timber production, which could be detrimental to shipbuilding interests. It is remarkable that Yarranton points to the possible replacement of wood by coal, which was not yet technically successful for smelting and refining iron. In places where coal but no iron works existed, firewood could not compete in terms of costs, with the result that the use of land for forestry was given up.

If it were not that there could be Money had for these Woods by the Owners from the Iron Masters, all these Copices would be stocked up, and turned into Tillage and Pasture, and so there would bei neither Wood nor Timber in these places: And the Reason is, Pit Coal in all these places, considering the duration and cheapness thereof, is not so chargeable to the owner of the Woods as cutting and carrying the Woods home to his House (ibid., 60).

Therefore, pure wage costs were higher in furnishing firewood than in using coal, even apart from the fact that the land could bear a rent. It must be considered that Yarranton's pamphleteering interest exaggerates in favour of the iron industry. However, from an economic point of view the price for firewood had to be so high that the land on which the cutting wood stood yielded at least the same rent as its next best alternative. If the price of firewood was very high over a longer period so that the rent to forest owners was permanently above that for pasture or arable, reforestation would be encouraged. Since firewood can already be cut after 12–20 years in coppices, the loss of rent for these years would have to be considered. Inversely, there are clearing costs in transforming cutting woods into pasture or arable, that in turn delay an adaptation to market conditions.

If we ask whether England experienced a wood shortage in the sense of an energy shortage, the question amounts to asking if sufficient wood was available in locations that satisfied other energy requirements. In the iron industry this meant that sufficient water power for bellows and hammers and sufficient transport possibilities on water existed so iron ore and charcoal could be brought in and finished products could be sent to their markets. If there was an insufficient number of locations that met these requirements, then a scarcity of solar energy existed in England. It did not matter if there was forest in inaccessible areas that was economically useless in terms of costs, or ecologically in terms of the energy balance. A clear indication that there was a scarcity of wood in this sense may be seen in the fact that the British iron industry quickly expanded at the moment when a more accessible, i.e. less expensive, energy provider could be obtained and employed – coal.

The breakthrough in iron smelting

The explosive growth of British iron production that began in the late 18th century was undeniably based on coal. As already mentioned, a process for smelting iron based on coke was developed by Abraham Darby in Coalbrookdale by 1710, but it only established itself in the last third of the century with a dizzying acceleration. How can this delay by 50 years be explained? Is it not a serious argument against the thesis that the stagnation of the British iron industry was attributable to a lack of fuel? (Flinn 1958, 1967). Three reasons have been cited for this delay since Ashton (1924; also Schubert 1957):

i.   The Darbys kept their procedure secret.

ii. A coal of particularly low sulphur content was available to them in Coalbrookdale. Without it their process could not be imitated.

iii. Pig iron molten by their process was of inferior quality.

According to Hyde (1977), whose argument I am following here, objections may be raised against these three points:

i.   The process could not be kept secret over such a long period of time, especially since workers switched from Coalbrookdale to other plants. There are indications that the Darbys explained their procedure to other iron producers. They had partners and shares in other companies that did not use this process.

ii. It is true that the coal of Coalbrookdale was particularly low in sulphur and, therefore, better suited for iron production than other types of coal. Similar coal was elsewhere only found in deeper layers and was not accessible with the customary mining process of the 18th century. However, this was also true after 1750, when other iron works began to use coal.

iii. It was technically possible not later than the 1730s to process pig iron smelted with coke into wrought iron. The phosphorus and sulphur content of coke-smelted pig iron was not higher than when charcoal was used. However, there was a higher silicate content, which increased the costs of processing it to wrought iron, but the quality of the finished product was the same.

The quality difference of iron smelted with charcoal and coke was expressed in differential prices. Charcoal iron, which was qualitatively better, was more expensive than inferior coke iron. Therefore, if coke iron became cheaper so that the increased refining cost was compensated, this would lead to a procedural breakthrough. Apparently this happened after 1780. As long as the cost differential when using coke was not large enough to balance price differences, there was no reason to use coke for smelting. In other words, it had to be considerably less expensive to use coke for smelting if the lower prices which resulted from the lower quality of pig iron were to be compensated.

Hyde shows that before 1750 the cost advantages were insufficient to balance for this price differential and, therefore, offered a plausible explanation for the delay in the dissemination of the coke smelting process. For this purpose he takes a random sample of 18 from a total of 70 charcoal blast furnaces. Two thirds of them had variable costs of less than £5 per ton of pig iron, only three were above £5/10/–. By contrast, Coalbrookdale spent c. £7. Therefore, the charcoal furnaces had a clear cost advantage. At the same time the costs of processing pig iron to wrought iron were higher by £1–£2 if the pig iron was produced with coke. To remain competitive, it should have been sold to the smithies for this amount less. Why did Darby use coke at all under these unfavourable conditions?

The reason is that he had found a specific market niche for his pig iron: the high silicate content made coke pig more liquid, which resulted in a finer cast. It was possible to cast thin-walled vessels that way, which only had half the weight of vessels cast from charcoal iron and, therefore, obtained a higher price (Hyde 1977, 40). The other iron works could not imitate him because Darby kept his casting process secret – as opposed to his smelting process. The famous breakthrough that Darby had supposedly made in the use of coal (or coke) in iron smelting was restricted to a relatively small field of application. However, cast iron could now be used in many areas that formerly depended on forged iron. The following cast iron goods were made in Coalbrookdale: pots, frying pans, mortars, pans for soap boiling, chimney plates, baking ovens, pipes. After 1724 equipment parts such as cylinders, pistons and pipes for Newcomen steam pumps were produced. It should be mentioned that Darby was a pacifist and so refused to manufacture weapons, which limited the application of his process to peaceful purposes.

The import of Swedish iron increased further over the course of the 18th century. Around 1750 only 43% of the wrought iron used in England was manufactured in the country. Due to the high import tariffs on foreign iron (25%–30% of the price), the British iron industry remained competitive despite high fuel costs. It also profited from the high transport costs for imported iron, particularly in the interior.

Pig iron smelted with coke was qualitatively inferior to pig iron smelted with charcoal in most applications, so its price had to be lower because processing into wrought iron was expensive. At the same time production costs were higher, so it could not compete with charcoal iron except in the aforementioned cast iron market niche. To escape this dead end, the production costs for coke iron had to drop below those for charcoal iron. A number of options existed: there could be technological innovations in smelting with coke, charcoal prices could rise or coal or coke prices might drop.

After 1750 all these conditions were met. First there was a general increase in wood and charcoal prices because land was being used more intensively and woodlands were subjected to increasing competitive pressures by alternative land uses. But there were no technological innovations in smelting iron with charcoal that would have buffered increased costs, so overall prices for charcoal iron rose. This development was contrasted by a lowering of production costs for coke pig. Technological innovations were the reason. On the one hand a new process for coking coal was developed (Morton 1966), at the same time the cost of coal itself dropped because production methods in mining improved. An opposing motion set in: rising charcoal prices and dropping coke prices, technical stagnation in smelting with charcoal and rising efficiency in the coke iron sector. Coke could bear a much higher weight of ore in the blast furnace before it crumbled. Coke blast furnaces could be made larger so that economies of scale became effective, such as the fixed costs for masonry. Charcoal technology had reached the apex of its development, while an innovative thrust occurred in the coke and coal sector that now favoured a quick dissemination of the process.

From the perspective of fuel supply, iron smelting was freed of the restraints set by a territorially defined supply of wood. There was also a decisive breakthrough in mechanical energy. Beyond a particular size it was not possible to operate bellows with water power. Furthermore, watercourse capacity became insufficient to meet the many demands that were now made on water as a driving force. In the late 18th century steam engines were increasingly employed for these tasks. About 1790 only twelve of the 83 blast furnaces that used coke operated without a steam engine. There should be no doubt that the steam engine accelerated the establishment of smelting with coke. Iron works now became independent of seasonal fluctuations. Fuel costs were not as critical since the machine could be operated with coal waste that was unsuited for charging a blast furnace.

Steam engines and coke loosened the locational restrictions of iron smelters: they could ignore the wooded area that had set an upper limit with regard to energy and could concentrate several blast furnaces in one location, since they did not depend on the limited capacity of a watercourse. They still required water for the steam engine and as a transportation route, but running water was no longer required. Rather, one could go about building canals that linked centres of production and supply. The new locational qualification for an iron works was accessibility for its coal transports. Bellows now determined the upper technical limit of a blast furnace.

Before 1760	1788	1805
Under 700 t	925 t	1500 t

Table 17. Average size of blast furnaces

(Hyde 1977, 43)

In the mid-18th century, refining too faced the problem of rising charcoal prices. A series of efforts were made to replace charcoal by coke. There was a total of nine patents for supposedly successful processes between 1761 and 1784. The Wood brothers achieved the first breakthrough in the 1760s with the so-called potting process, which separated iron from coke with clay vessels ('pots'). This process prevented the transfer of undesirable substances to the wrought iron during heating and was very successful until the introduction of puddling in the 1780s. In the potting process, which half of the producers used around 1780, raw material costs, in particular fuel costs, lay significantly below conventional costs for the use of charcoal.

Finally, in 1783–84 Henry Cort developed the puddling process, which also separates pig iron from coke. This process permits a direct coupling of refining to rolling and therefore a new scale of production. The bottleneck that had continued to exist after the economic breakthrough of smelting with coke, because refining still required charcoal, was overcome. British production of wrought iron rose by 70% between 1750 and 1788; pig iron production increased in the same period by 150%. The switch of pig iron production to coke as its fuel was a precondition of the potting/puddle process because it used more pig iron than conventional refining process. This waste became economically possible because of less expensive coke pig. At the same time potting/puddling affected pig iron production because a larger market for inferior pig iron with high silicone content was created.

The breakthrough of British iron production towards the end of the 18th century was due to an energy transformation that occurred in two phases: first pig or cast iron was manufactured with coke, which of itself did not induce growth in iron production. Only when it had become independent of wood in refining as well did iron production expand on a grand scale. If one assumes, as was customary in the older literature on the history of technology, that the transition to coal as the energy source in the iron industry was already complete in principle by 1710, the slow diffusion of the process before 1780 speaks against an actual shortage of fuel. If the entire process is considered, it becomes apparent that it was more complex: the critical bottleneck was in refining, i.e. the manufacture of wrought iron. Only when the technical problems associated with that were successfully resolved did a complete change of the energy basis of heavy industry become possible.

The first result of the new process was a rapid decrease in iron imports from Sweden and Russia – with the exception of high quality Sweden steel, which continued to be produced with charcoal. This link became apparent in a mocking song that the workers in the iron works of John Wilkinson sang (cited by Ashton 1924, 87):

> That the wood of old England would fail did appear,
> And tough iron was scarce, because charcoal was dear,
> By puddling and stamping he prevented that evil,
> So the Swedes and the Russians may go to the devil.

A temporal resource expansion emerged instead of a territorial one. The geological past of photosynthesis was accessed and liberated from dependency on the annual increase of wood and associated territorial restrictions. The ecosystem of the English economy no longer extended to northeastern Europe but back into the Carbonaceous period. What this meant for iron production may be demonstrated by contrasting smelting based on charcoal and coke (Table 18).

As we can see, the number of possible locations multiplied. At the same time a concentration of producers in the same location became possible since the elements of production, ore and coal, could be transported almost at will. This was impossible over longer distances for both charcoal and water-produced mechanical energy. Blast furnaces and foundries could move together and the basis for a concentration of heavy industry was created. The transition to a coal economy had far-reaching consequences for the organisation of the iron

---

1. Capacity

Smelting (per blast furnace)

| 1720 | ~ 300 t/year | max. 700 t/year |
| 1815 | ~ 1,500 t/year | max. 3,000 t/year |

Refining (per foundry)

| 1720 | ~ 150 t/year | max. 350 t/year |
| 1815 | ~ 5,000 t/year | max. 13,000 t/year |

2. Capital size

Beginning of 18th century	£4,000 per blast furnace
	£1,500 per foundry
Beginning of 19th century	£20,000 per blast furnace
	£80,000 per foundry

3. Power source

Wood:　water for bellows and hammer works, with seasonal disruptions.

Coke:　steam for bellows, rolls and winches for charging blast furnaces, continuously available.

4. Transportation

Wood:　pack horses, carts, tow boats

Coal:　horse tracks with rails, canals, then trains

5. Maximum size

Wood: 2,000t/year in one location

Coke:　no clear limit

6. Locational requirements

Wood: ore, water gradient, wood, water route

Coke:　ore, coal, water route, later railway link

Table 18. Comparison of wood and coal/coke in ironmaking

industry. Nothing stood in the way of an expansion in size on the supply side. At the same time the existing narrow locational requirements disappeared. The typical concern of the new era was integrated in its structure: it mined the ore and coal, smelted, refined, rolled, and slit the iron into its finished form of plates and rods. From this time onward, therefore, very large capitals began to be sunk in the industry' (Ashton 1924, 100). This new branch of industry now concentrated in the proximity of the coal fields of the Midlands, Yorkshire, Derbyshire and South Wales. The accessibility of coal became the decisive locational requirement.

If one wishes to speculate what course European development might have taken if the transition to a fossil energy system had not taken place – perhaps because there was no coal or because it only occurred in inaccessible locations – then it can be said with some certainty that the tendency would rather have run towards decentralisation, deceleration of development and finally a stationary state. This tendency towards deceleration would have been effective not only on a local scale but even more so in larger economic regions. Without doubt there would have been no Ruhr district and no Black Country, but iron industries in Sweden, the Urals, Asturia, the Siegerland, Carinthia and similar decentralised locations would have retained their old importance.

However, if the social and cultural development in the spirit of unfolding capitalism had expressed its peculiar dynamic independently of the fact that a huge energy reserve rested in the fossil reserves which permitted the material realisation of this dynamic on the scale we know, then even without coal a series of technological and economic innovations could be anticipated. Instead of an Industrial Revolution with its characteristic acceleration of several technical, economic and finally political and cultural processes there might have been an industrial evolution. European societies would have remained more decentralised in economic and settlement structure and developments would have occurred significantly more slowly. One may speculate whether European economies would have quickly entered into a stationary state or, alternatively, whether improvements in the use of plant biomass, discoveries in energy-neutral areas and improved efficiencies would have resulted in long-term, unnoticeable growth. If population development had taken its familiar course without the industrial leaps in growth of the 19th century, that society would have faced the problem of increased struggles over the distribution of scarce resources that will only arise in the 21st century.

With the transition to the use of coal the energy framework of society changed, which in turn had economic and ecological repercussions. It became clear that the fossil energy system initiated emancipation from restrictions imposed by the agrarian solar energy system. The agrarian system was in the end tied to the territory from which it derived its energy basis. Fossil energy liberated humans from their ties to area size. We have seen that Britain already in the first decades of the 19th century disposed of a quantity of energy that was equivalent to a hypothetical area the size of the entire country. But with that development the territorial solar energy principle was broken.

A comparable consideration may be made for the iron smelting sector. Under preindustrial conditions about 50 m³ wood were required to make 1 ton of wrought iron, which was equivalent to the sustainable yield of 10 ha of forest. On this basis it is possible to calculate the hypothetical area that was necessary to obtain the known historical production figures of iron with charcoal.

1620	1.9
1720	2.5
1750	3
1781/90	6.9
1791/1800	12.7
1815/19	33
1835/39	114.2
1840/44	146
1855/59	358
1875/79	648
1895/99	877
1910/13	979

Table 19. Territorial equivalents of British iron production
(1000 km²)

If these figures are related to Britain's total area of c. 228 km², it is clear that iron production up to the mid 18th century moved within the frame of what was theoretically possible on native territory in terms of energy. About 1750 not even 2% of the land area was used for supplying the iron sector with wood. At the beginning of the 19th century, it was already 10%, in the 1840s the territorial capacity of the country was exceeded and by the First World War Britain would have required four times its territory if its iron production had relied on charcoal alone – though our figures use the technological state of the 18th century and consider no innovations in smelting with charcoal, which would undoubtedly have been possible.

But even if technological improvements in fuel consumption efficiency are considered, it is questionable whether in the light of locational restrictions (apart from a supply of wood and water power this includes suitable transportation routes) the entire European iron production would ever have reached the state of Britain's at the beginning of the 19th century. A society without coal would have remained a wood-dependent society not only with regard to the supply of firewood. We may assume that analogously to English iron production in the early 18th century, European iron production would have soon reached a plateau with strongly decentralised distribution. It would surely have surpassed the state of 1700 but hardly much that of 1800. This would not have meant that Europeans would have lived in the Stone Age, but that their world would not have been as heavily cast in iron.

According to Braudel (1974, 282) the entire European iron production in the early 16th century stood at about 100,000 tons a year. This is a very crude estimate in which figures from regions for which detailed information is available are averaged out for the entire territory. (Basque country 15,000 tons, Styria 8,000–9,000 tons, Liège 10,000 tons, France 10,000 tons, England 6,000 tons, other German territories 30,000 tons. Figures for Sweden, Russia and Asturia are not available). In the same period the entire European population (including Russia) may have numbered about 100 million. That would mean that 1 kg of iron was produced per capita in the early modern period including military demand. That was not much, even if the figure were doubled. No wonder that iron was scarce and expensive.

Around 1700, British iron production stood at 20,000 tons per year, while the population was less than 10 million. That was just 2 kg iron per person. However, the same amount again was imported. By 1800 the entire population had grown by about two thirds while iron production had grown twelve fold to 250,000 tons a year. Now 15 kg of iron were available per person, which obviously lay above the margin of what was possible in agrarian societies. (A comparison: world production of iron in 1990 was 785 million ton, while population stood at 5,300 million. That was after all 150 kg/person annually, maybe more than one hundred times what was possible under the conditions of an agrarian solar energy system).

In the middle of the 18th century, 30–50% of British iron demand emanated from agriculture (Bairoch 1978, 491). The great push in improvements of agricultural productivity that set in around 1700 rested among other factors on a series of technological improvements that were linked to a high demand for iron: wheels with tyres on wagons, shoes for horses and cattle, ploughshares, harrows, rakes, spades. In particular, the use of iron for agricultural tools had as an effect an increase in efficiency in mechanical work. An iron plough is easier to draw than a wooden plough, so instead of three or four draught animals only two needed to be employed. Also, an iron tool was more durable than its wooden counterpart, which effected further savings. In the context of the solar energy system, this was a typical substitution of land uses: iron instead of draught power meant wood instead of pasture. If iron is produced with coal, a net gain in area occurs.

The 'agricultural revolution' of the 18th century (Chambers and Mingay 1966) made it possible to feed a rapidly growing population. If one assumes that European iron production had already achieved its sustainable maximum in the 18th century, it can be surmised that a significant amount of iron was employed in agriculture. Moreover, the iron demand of the military sector must not be forgotten. In contrast to agriculture, in warfare there is a rapid and irreversible iron consumption. In agriculture iron tools (or iron parts on wooden instruments such as ploughshares or spades) were employed that lasted until they were corroded. The iron ammunition of muskets and artillery was quickly used up in a battle and could not be recycled. If the enormous growth of armies in the 17th and 18th centuries is considered, it is clear that a large part of the increased iron production must have been used for weapons and ammunition.

In this light it seems clear that iron would never have penetrated all areas of life within the framework of the agrarian system as it did after the British breakthroughs with coke at the end of the 18th century. The iron industry formed the base of heavy industry in the Industrial Revolution. Its growth was the material precondition for an expansion of machine building in leaps and bounds. Smokestacks are not the symbol of an age that emphasised coal and steel without reason. However, it is inconceivable that there would have been a development worthy of the name Industrial Revolution based on charcoal.

3. Transport and Steam Power

Transportation of coal

Apart from breakthroughs in iron smelting, the establishment of the new fossil energy system started other processes, of which the transformation of the transport system will be considered in more detail. As has been noted above the accessibility of an energy carrier is directly dependent on whether the quantity of energy required to transport it is larger or smaller than what it makes available at its destination. It was been calculated that, in terms of energy, a ton of coal is equivalent to the annual yield of a hectare of forest. Obviously, it is much easier to transport a ton of coal from the pit to the consumer than to collect five cubic metres of felled wood from a hectare and transport it on forest roads in horse carts (see Wrigley 1962).

Let us consider in some detail the structure of spatial distribution of wood and coal. We may assume that 1,000 m³ of timber stand on a hectare of forest, which is the case if it is an approximately 200-year-old forest. Let us also assume this volume of wood was evenly distributed over this area of 1 ha (= 10,000 m²). The forest floor would be covered with a 10 cm thick layer of wood. Since the combustion value of wood is approximately 30% of coal, an equivalent layer of coal would have a thickness of 3 cm. In terms of energy a 200-year-old forest is equivalent to an area of equal size covered with a 3 cm thick layer of coal.

However, mineral coal lies underground in deposits of up to 30 metres thick, with 30–120 deposits possibly lying on top of each other. Let us assume that the percentage of coal in the entire mining area is only 10%, in which case 100,000 m³ coal would be contained in a cube with a side length of 100 metres (i.e. with a surface of 1 ha or a volume of 1 million m³). On the other hand, for 1 ha of forest the coal equivalent is only 300 m³. From this calculation it is evident that the energy potential of a certain area in coal mining is at least 300 times higher than that of forestry. These geometric properties explain why the removal of wood must be more expensive than transporting coal.

The transportation of coal begins concentrated at one spot. It is brought from its deposit to the mine exit from where it is transported to the consumer. The transportation route remains the same for a long period of time and a great quantity of coal, so that it is worthwhile to build relatively costly facilities such as a railway. The expense that is incurred for a any particular quantity of energy declines. The heating value of mineral coal is 30–35 MJ/kg, but of firewood only 12–15 MJ/kg; the density of coal is 1.2–1.5 and that of wood 0.4–0.7. For a particular amount of energy, only a seventh or eighth of the freight volume is required for coal compared with firewood. Therefore, transport over large distances is worthwhile.

In the early stage of coal mining no special transport systems were created. Coal mining was adapted to existing transport facilities. First, coal occurrences lying directly on watercourses were mined so that coal could be loaded directly from the mine shaft into small boats. The cost of transport to reach the consumer was critical for establishing a mine. For a number of reasons, especially drainage, it was more economical to sink new shafts again and again than to extend existing ones. Until the 18th century shafts rarely went deeper than 200 metres and even there considerable difficulties arose. Many pits were no deeper than 8–12 metres. There were exceptions: thus, a mine was reported in the 17th century that was driven one mile under the sea. No transportation costs arose from the mine exit to the ship, so that the additional effort underground was apparently worthwhile (Nef 1932, 1, 367f.).

Mining methods were very inefficient. Large quantities of coal were left underground to support the shafts. Only what could be easily accessed without much danger was produced. Techniques for supporting the shafts were barely developed. Since wood for the pillars was expensive, it was usual to leave columns of coal and in this way, too, a substitution of wood by coal took place. Until the 18th century transport methods within the mines were very primitive. Coal was carried in baskets to the shaft or pulled on sleighs and hoisted up with a manual winch – in exceptional cases also by a winch that was powered by a horse gin. Animals were not used underground until the 18th century. Usually it was less expensive to sink a new shaft than to extend the tunnels underground. The quantity produced per shaft was not large on average. In the 17th century it lay between 3,000 tons and a maximum 20,000 tons annually. A pit was usually operated for no longer than two years and new shafts were opened all the time.

Two possibilities remained when deposits immediately next to watercourses were exhausted: they could be extended or new pits could be opened in the interior. One option incurred higher transport costs; the other required that pit draining methods were developed. Land transport required pack horses, which had only a very limited carrying capacity, or horse carts, which could not handle unpaved roads in the long run. It was calculated that the coal price doubled every two miles in overland transport (Nef 1932, 1, 103). Wooden railways were built from the 17th century for transport from the pit to the water. These were laid out so that the loaded wagons ran into the valley driven by their own weight and manned only by a brake operator. After emptying, horses pulled it back up again. To achieve a constant but not too steep incline from the shaft exit to the ship-loading dock, bridges, terracing and levelling had to be carried out. The associated considerable capital investment was only worthwhile if larger quantities over some time were guaranteed. The exploitable reserves should not be too scant and there was a premium associated with prolonging mining at one location through investments underground. Rails shortened transport times and therefore increased route capacities. The amount produced could be increased within a given period, which decreased the turnover time for capital. Furthermore, transport in rainy weather then became possible – normal roads were only suitable for transport in a dry or frozen state.

These railways were direct precursors of iron railways. First, wooden rails were laid as a plank way for rolling wagons. Then wheels were equipped with tyres and rails with a profile so that the wagons would run evenly and did not have to be held in the track. Since wear and tear on the rails resulted in great consumption of wood, they were lined with iron bands; and in the middle of the 18th century experiments with cast iron were undertaken but the rails broke easily under the heavy weight of the coal carts. After the development of the rolling process in connection with puddling, more elastic rails were available that finally made it possible to put a heavy steam engine on wheels to draw coal carts.

Transportation followed watercourses after the coal had arrived at the loading site. The greater part of transportation was by water. Nef provides a particular fitting example:

To reach the town of York, Newcastle coal had to be carried at least two hundred miles by water, and loaded five separate times – at the pithead into carts, at the wharves from carts into keels, from the keels into sea-going colliers anchored in the Tyne, from the colliers into lighters on the Humber, and from the lighters into carts at York for carriage to the house of the consumer or merchant. Yet, until the deepening of the Aire and Calder at the beginning of the 18th century, York received some of its coal from the north of England, in spite of the fact that supplies could be had by land carriage from the mines in the West Riding only twenty miles away (Nef 1932, 1, 102).

Despite the costs that were incurred by reloading five times en route from Newcastle, its coal competed with Yorkshire coal in York solely because of transport by water. This clearly shows how critical the transportation element was for coal supplies. They were cheaper in places that were accessible by water than in the interior. The expansion of the coal economy created a tremendous demand for transport capacity and provided the impetus for building special ships. English coal was even exported to the Continent. However, in this case transport dedicated only to coal would not have been worthwhile, given the ratio of value to weight. It was used as ballast in the transport of light goods that could be sold when heavy goods were loaded.

Given the obvious advantages of waterborne transport, building of canals from coal mines to ports and centres of consumption began in the 18th century. According to a report of the year 1803, 165 laws governing the construction of canals (Canal Acts) were passed by Parliament in the preceding 45 years. Of these, 90 served coal mines and 47 ore mines (Dyos and Aldcroft 1974, 111). Usually these canals were privately funded and paid for by user fees. In transporting heavy goods they were far superior to roads and were only pushed out of the market by railways in the 19th century.

Power sources for draining coal pits

Apart from the transportation problem there were other fundamental difficulties. When shafts reached a particular depth, groundwater collected and prevented continuation of work. As in ore mining, a remedy was initially provided by digging a drainage shaft. However, the precondition was that the exit of the drainage canal had to lie below the lowest point of the shaft: the shaft had to be in an elevation and the water had to drain into a valley. The number of possible shafts and their furthest depth were strictly limited by this condition.

The first escape from this difficulty, draining devices after the model of Continental ore mining began to be used in the 17th century. These were rather complicated systems that brought water to the surface with chains of pails or pumps but they required a high energy input. To provide the driving force the three familiar sources of mechanical power were used – wind, water and conversion by animals.

A water mill was the least expensive and most reliable source of mechanical energy, but it had the serious disadvantage that it could only operate on a watercourse. Its mechanical energy could only be transported over relatively short distances of at most 1000 metres using wheels, rods and belts, because friction losses were too great. Initial construction required rather large investments for the building of reservoirs and dams, of the mill itself and, finally, for the transmission to the pit. In return maintenance costs were low. A serious problem was that watercourses might freeze in winter and dry up in summer. Therefore, continuous use was much restricted. This problem affected not only coal mining but all operations that depended on a mechanical power source. The peculiarity of coal mining was that the shaft exit, as opposed to a spinning mill, could not be positioned at will.

A windmill's advantage as a power source is its independence of location. It could be employed inexpensively anywhere where there was wind, as was the case in many places. It was also inexpensive to set up, though its operation was more difficult than that of a water mill because it always had to be turned into the wind. Its main disadvantage lay in the fact that it could not be continuously in operation but depended on medium wind strength. If the wind was too light or too strong it could not be used. Allowance had to be made for downtimes of 75%. In mining this could mean that the pit drowned during a longer down period. There was no other possibility for storing the energy it delivered during operation except building an extensive water reservoir to power a waterwheel during a calm or a storm. Such a device was only sensible in combination with a water mill, and the windmill was only used in that case to increase the useful quantity of water.

This was only reasonable in situations with a permanent water shortage, for example on small rivers that were heavily used and in which refilling was not guaranteed over the entire day. Since there was always a risk that a mill would dam water upstream that would then be lacking downstream, it was sensible to keep a certain amount of water in a reservoir on one's own account. There were indeed rivers where water power was used to the fullest. Thus, the Irwell has a gradient of about 300 metres in total. No fewer than 300 manufacturies with their own mill were located on it, so that there was an average drop of one metre (Hills 1970, 95). The entire energy available to a water-powered device consists of the product of the quantity of water and the height of the drop. It therefore made sense to use all the water that flowed by in such a way that all energy that passed by in the day was used during operation times of 12–14 hours. This was achieved with reservoirs that served as intermediate storages. If such a system was coupled with a windmill, its low reliability was only of subordinate importance and additional costs were not excessive.

The third source of mechanical energy was draught animals, especially donkeys, oxen and horses, which walked on a horse gin. They had the great advantage of reliability and continuity but they were the most expensive. Including the wages of the driver, gin operation over a year cost five times as much as a windmill and three times as much as a water mill per horsepower delivered. But there was no choice in certain locations, if a usable watercourse was unavailable. Thus, 50 horses were employed for a mine in Bedworth, Warwickshire, to operate water-raising machinery, though this was obviously an exception (see Nef 1932, 1, 375). Landes (1969, 98) mentioned a mine in Warwickshire that supposedly employed 500 horses but one may suspect this is an error by an order of magnitude.

These three sources of power were used to operate water-raising machinery that usually consisted of a series of buckets. These devices were very inefficient. A large part of the water flowed back into the pit during transport. Furthermore, the coarse wooden water-raising machinery was itself heavy and its weight increased relative to the water raised with increasing depth, so that there was an upper limit for such water-raising chains. The number of work animals could be increased or the material could be made lighter but the expenses, downtimes and costs rose. At most four animals could walk on a horse gin and even then difficult problems arose in the co-ordination of force transfer. Some very elaborate systems were in use that raised water several storeys over intermediate basins and brought it to the surface there, but the costs were enormous. Such techniques were developed in silver ore mining and could only be transferred to a limited extent to coal mining.

From the steam pump to the rotating steam engine

In the 17th century a technical method was sought to pump water more reliably and effectively than was the case with conventional water-raising devices. The first breakthrough was the atmospheric steam pump, which Thomas Savery presented to the public in 1702. However, it had the significant disadvantage that it produced little suction and therefore had to be installed about ten metres above the water level. A deeper mine required five or six machines of this type to pump the water up in stages. Furthermore, this machine was unsafe and exploded easily.

Thomas Newcomen built the first really useful steam engine on the basis of Savery's patent. A prototype was installed in 1712 and the use of the machine rapidly spread. A contemporary wrote that the Newcomen steam engine could raise as much water as 50 horses, which would cost at least £900 annually, while the engine would only cost £150 annually in coal, maintenance and repairs (Galloway 1882, 82).

This machine worked according to the following principle: the steam was condensed in an open cylinder in which there was a piston. The suction created pulled down the piston. It was attached to a balance-like construction at the end of which another piston was attached that was pulled up. It extended into a pipe and created negative pressure, which drew up the water. Afterwards steam flowed again into the condensation cylinder and the two linked pistons returned to their original position. The machine had a very slow rhythm, at most 12 movements a minute. That was the reason why no rotation could be produced with it. Despite its huge size it was well suited for use in mines.

Its energy efficiency was very low, probably below 1%. This tremendous waste of energy was due to the fact that the steam in the condensation cylinder was cooled after every beat and lost. Also, the iron parts were manufactured with little precision and the pistons had to be tightened against the cylinder walls with leather fittings. Even so, the use of this machine was economic because the fuel cost almost nothing in a coal mine. The steam pump was fired with coal waste, dust and small bits that were unsellable and would have remained in the waste heap. One may ponder whether under other conditions, without a free source of energy, the initial hurdle for the steam engine, which was its enormous requirement for energy, could have been overcome at all. On the other hand, the low price of energy was the reason why the significantly improved Boulton and Watt steam engine only established itself fairly late in coal mining, although it consumed far less fuel. It was more expensive – in part because of the high patent costs – than the Newcomen steam pump and also more complicated and more dangerous.

Steam engines of the Newcomen type quickly spread from coal mining into other fields dependent on mechanical energy, such as the textile sector. Where rotational motion was required, it could not be employed directly, but in connection with a water mill it provided good service as a pump. It returned the water that had operated the mill wheel back to the reservoir. In this way a manufacture could largely become independent of a natural water flow but coal was required. For this reason the steam pump was often only kept in reserve and work was usually performed by natural water power – water flowed for free but coal cost money. This system was maintained until the 19th century, not only for reasons of cost but for reasons of quality. In particular the textile industry was dependent on an even motion without any jerking because the threads tore easily. The Boulton and Watt rotating steam engine was not always capable of fulfilling this requirement, despite its flywheel. The speed of revolution of a mill wheel was easier to regulate than that of a steam engine. Furthermore repairing the steam engine caused problems compared to the robust wooden water wheel.

If the steam engine went down, it usually did so abruptly and disrupted production. The mill wheel could be repaired during the times that the business cycle permitted.

Towards the end of the 18th century there were a number of improvements in steam engine construction. James Watt's patent for a more economical machine with a separate condenser originated in 1769 but it was only ready for production around 1780. At the same time James Pickard invented a process that merged the up and down motions into one rotation, but without a flywheel or regulating valve. Around 1785 the Boulton and Watt rotating steam engine was in principle ready to be employed. In 1788 the automatic centrifugal governor was invented. The problem was that the speed of the rotation was dependent on stroke distances and they in turn on steam pressure. Attempting to maintain as even a fire as possible never succeeded with the exactness with which the flow of water over a mill wheel was controlled. The invention of the centrifugal governor, which made automatic control of the rotating steam engine through feedback possible, was a major precondition for the use of the steam engine in the textile industry and other trades.

The Boulton and Watt steam engine was employed parallel to the Newcomen type of steam engine till 1800. In that year James Watt's patent expired and the fuel savings of his machine quickly resulted in its general success. Prior to this, high patent costs had for the most part cancelled out savings in coal. After 1800 steam engines were increasingly used in iron smelting, especially in bellows operation for blast furnaces but also in hammer mills and rolling in foundries. Until 1790 smelting work was mainly carried out with water power but steam bellows established themselves quickly for two reasons. For one, locations with sufficient water power and sufficiently close to raw material sources and markets became increasingly scarce. Furthermore, new demands were made on power supply. Traditional bellows could not deliver the performance required for the new, large blast furnaces. A suitable piston bellows was more easily operated with a steam engine than a water mill.

The Boulton and Watt rotating steam engine was readily employed for transport as well. As early as 1807 the American Robert Fulton built the first steam boat, and the improved high pressure engine of Oliver Evans made it possible to reduce the size of the machine sufficiently to set it on a wagon. Rotating steam engines could easily be used in transportation. Steam locomotives were first employed after 1815 in coal districts, since fuel was inexpensive there, but not fodder for horses. The victory of trains as a universal means of transport was clearly favoured by the Corn Laws of 1815: import of grain was made much more expensive by tariffs so that domestic price levels remained very high. This resulted in an increase of the cultivated area so that the costs of keeping horses increased. Pastures now competed increasingly with arable. This increased the incentive to replace the transport system of horse/hay/land with the system of steam locomotive/coal/mining.

This substitution effect was presented as the decisive argument in the debate regarding the introduction of the railway. It should not be forgotten that the public discussion of the 1820s and early 1830s was marked by a fear of overpopulation, fanned particularly by Thomas Robert Malthus and his adherents. If population grew and land was scarce, every innovation that made additional land available had to be a blessing. When in 1833, a year before the reform of the British Poor Law, a committee of the House of Commons was to decide whether the use of steam engines and rails should be expanded, an expert testimony emphasised the possible socio-political importance of a transition to fossil energy. Transition to coal could defuse the problem of poverty that had arisen because of population growth:

> It has been said that in Great Britain there are above a million of horses engaged in various ways in the transport of passengers and goods, and that to support each horse requires as much land as would upon average support eight men. If this quantity of animal power were displaced by steam-engines, and the means of transport drawn from the bowels of the earth, instead of being raised upon its surface, then, supposing the above calculation correct, as much land would become available for the support of human beings as would suffice for an additional population of eight millions; or, what amounts to the same, would increase the means of support of the present population by about one third of the present available means. The land which now supports horses for transport or turnpike roads would then support men, or produce corn for food, and the horses return to agricultural pursuits (Cundy 1833, 21).

This argument was convincing, and true in principle, because an area that fed a horse could yield plant food for eight humans. However, it would be shown that the railway would not displace the horse by any means. Since transport volume increased considerably with the railway, but the system was only laid across the land as an open network, the requirement for transport services across short distances rose dramatically. In Germany the number of stagecoach passengers increased between 1840 and 1873 from 3 to 7.4 million and even after the turn of the century it lay at about 4.5 million. An indication of the increase of commercial transport of goods with horse carts may be seen in the fact that the number of persons employed in this sector in Prussia doubled between 1846 and 1895 (Sombart 1913, 248f.). The number of horses rose in Germany from 2.7 million at the beginning of the 19th century to 4.6 million in 1913, which also included animals employed in agriculture (Henning 1978, 2, 26f). Only the massive spread of the automobile and the tractor during the 20th century finally pushed the horse into the leisure sector.

Coal mining created demand for the steam engine. Its use permitted expansion of the latter that would not have been possible with conventional means. The establishment of steam engines in turn increased the demand for coal. It was calculated that in 1800 alone steam engines of the Boulton and Watt type installed in Britain used approximately 270,000 tons of coal altogether (Tunzelmann 1978, 111). That was about 3.25% of British coal production that year. If we take all machines, even the wasteful ones of the Newcomen type, steam engines probably accounted for the consumption of about 10% of coal production. If we assume that about two thirds of total coal production was used for home heating, a third of commercial consumption was due to steam engines. Their share in overall coal consumption rose tremendously during the 19th century. In the middle of the century an eighth to a sixth of British coal production was probably used to fire steam engines. If consumption by locomotives and steam boats, which are not contained in these figures, is added, together with consumption during the production of iron for steam engines, railways including rails, bridges etc. and steam boats, the sum justifies the conclusion that the growth in British coal consumption from 10 million tons in 1800 to 60 million tons in 1850 may be attributed in large part to the steam engine complex.

Through a counterfactual calculation Tunzelmann (1978) came to the conclusion that the steam engine could have been replaced before 1800 by other energy converters, especially water power, without major economic consequences. According to his calculation the gain in wealth by employing the steam engine was only 0.2% of the national income. In other words, the national income without steam engines would have been smaller by this fraction. The problem in his calculation is that Tunzelmann calculates in money alone and does not consider material flows and natural factors. Otherwise, he would have had to ask to what degree the use of water power could have been expanded, how the bottlenecks in draining coal pits would have been overcome and how the technical problems in iron smelting would have been solved.

4. Significance of Coal in the Industrial Revolution

The final breakthrough of coal as the new fuel in the late 18th century originated in the iron industry and coal mining itself. The invention and spread of the steam engine and innovations in transport leading up to the building of the railways originated in those two sectors. The steam engine then spread to other branches of industry but there the introduction of the rotating steam engine often stemmed not from an actual problem in energy provision but merely from slight cost advantages over water power that were not always obvious. Whether costs really were lower depended on a number of factors, among them purely technical and geographical ones. Water power was as a rule cheaper in places where coal was expensive, that is in remote areas far away from deposits and with unfavourable transportation. In other areas the potential of water power was virtually exhausted and rents were high at suitable locations. The spread of the steam engine lasted well into the 19th century, an indication that it was not always and everywhere superior to competing water power.

The energy and material basis of the textile industry remained solar for a long time. Its raw materials – wool and cotton – consisted of biomass and transport was primarily carried out with sailing ships, that is, using wind energy. The textile machinery itself (water-frame, spinning jenny, mule etc.) were almost entirely made of wood and therefore hardly required energy-intensive iron. The mechanical drives of spinning and weaving mills were primarily supplied by water mills until well into the 19th century. The mechanised textile industry was in its material and natural aspects not so much a pioneer of the industrial system as an outgrowth of the agrarian mode of production. The great advances and increases in productivity at the end of the 18th century and the beginning of the 19th century would have been possible in this sector without coal, iron or steam – at least if the invisible problem of alternative land uses in England is disregarded. This can only be approached through analytical observation and in any case is moot for imported cotton.

If a longer period of time had been involved, textile industry growth might have been expected to settle into a stationary state. Above a particular size and speed, textile machinery had to be made of iron, since wooden machines work with less precision. The problem of a lack of iron due to locational restrictions of iron smelters would have become critical at some point, i.e. iron prices would be likely to rise. Also, the amount of water energy available in a particular region could not be increased, so rents on watercourses would have risen rapidly. This might have induced progress in mill technology but it is questionable if the threshold to the electrical economy would ever have been passed, since it relied in material terms on the availability of large quantities of inexpensive metals. There was the possibility of substitution by horses (which require pasture) and in a limited sense by windmills, but this type of economy would have headed towards the state predicted by economists in the early 19th century: rents and land prices would be extremely high, wages low and economic development would stagnate.

Without the unbroken industrial dynamic the chemical industry, too, would soon have faced serious energy problems, especially in the production of soda, sugar, sulphuric acid and alcohol (Schmieding 1991). In particular the great breakthroughs in agricultural chemistry from the final years of the 19th century would hardly have been applicable in large-scale industry. Huge quantities of energy were required for ammonia synthesis in the Haber-Bosch process. Increases of agricultural yields linked to chemical fertilisation, pest control and mechanisation, which permitted an improved diet for the growing population of industrial countries, would have been impossible without fossil energy. The solar energy system must have a positive energy balance especially in agriculture, since there is no possibility of subsidising agriculture from other sources. Good caloric provision and a high component of meat in the diet today may be attributed in the first place to the fact that more energy is invested in agriculture in the form of fossil fuels than its biomass returns. Industrialisation without fossil energy would at least have meant that other trades would have had less energy available.

An important effect of the use of fossil fuels was the technical acceleration of many processes of production. Three weeks were required until completion to process pig iron into wrought iron using the preindustrial method of hearth refining with charcoal. At the end of the 18th century the puddling process with coke still took two and a half days. The Bessemer process developed in middle of the 19th century was complete after only 20 minutes (Sombart 1902, 75). Even if it had been possible to increase considerably the number of iron works scattered over the countryside, industrial development would have been much slower if the production time in this key sector could not have been shortened significantly.

Considering the central importance of coal as the energy basis of the Industrial Revolution, it is quite astonishing that, as Wrigley (1994) noted, it has been almost completely ignored by economic history. This disregard has a long history of its own and can be traced back to the tradition of classical political economy from Adam Smith to John Stuart Mill, who treated resource questions entirely as questions of agricultural production. Ever since the historical project of Karl Marx the textile industry has been considered the pioneering sector of the Industrial Revolution. The innovations that emanated from there particularly affected socio-economic structures: the factory system was to a large degree established in the textile sector, new behaviour patterns were generated and established that resulted in the dissemination of industrial work discipline. Textile workers were an initial focus of resistance against capitalism, so here the making of the working class and the labour movement could be studied. Nor should one underestimate how influential it was that one of the greatest socio-political writers of the 19th century, Friedrich Engels, could study the textile industry at first hand in Manchester.

From the perspective of social history in particular, coal mining can be viewed as a largely traditional affair. The factory system with its novel discipline cannot be found there and the new mechanised techniques only played a marginal role. Mining was manual work until the end of the 19th century. Pick and shovel dominated the work process and the personal expertise of miners was still crucial. This technical stagnation manifested itself in an absence of increases in labour productivity. Around 1700, coal production per miner in Britain was about 120–200 tons per year, and around 1900 it had risen to just 250–300 tons per year (Wrigley 1988, 77).

However, the energy yield of coal was considerable despite this low productivity. If a worker produced 250 tons of coal in a year, that was about 1,000 kg coal per day over 250 work days. The combustion value of coal is 30 MJ/kg, so a miner produced 30,000 MJ daily while consuming food with an energy content of 12 MJ. The output factor is no less than 2500 (not a mere 500, as calculated by Wrigley 1988, 77). Even if we relate the combustion value of annual production of 7,500,000 MJ to the annual dietary input of 4,380 MJ, we still arrive at a yield factor of 1,712, which lies well above the potential of agriculture and forestry. Perhaps this high productivity placed a low premium on technical progress in coal mining so that we face a strange paradox: the labour basis of the Industrial Revolution remained a conservative matter.

Thus, the technical and economic basis of the new energy system was not particularly spectacular. That may be the reason why the consequences were not systematically noted. However, they consisted of a dramatic alteration of the economy's energy framework: within the energy system there was a fundamental mobilisation of material flows that is readily apparent in the following diagram.

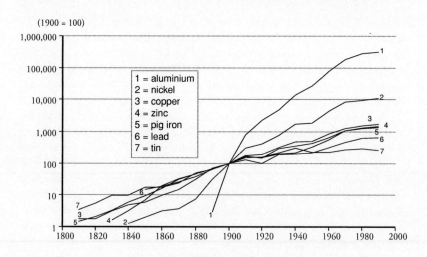

Figure 4. Global production of metals
(Sources: Rössing 1901; *Metallgesellschaft* 1992; *Historical Statistics of the US* 1976; United Nations, *Statistical Yearbooks*)

Energy expansion pulled other materials along with it. They were mobilised from their deposits, transformed, concentrated and finally distributed over large areas. The superabundance of fossil energy put metals into frenzied circulation. Metal compounds available in ores were first reduced but then formed new chemical bonds. This process is the metabolic basis of the new scale of the pollution problem as it arose during industrialisation. These substances were not only mobilised and distributed over the entire globe, they also reached unintended destinations in unpredicted toxic concentrations. Not only did fossil energy make access to material resources possible, it also created a tremendous requirement for sinks, in which to deposit these resources after use.

The history of energy is the secret history of industrialisation. Some observers have noted that the material basis of the economy was changed by industrialisation. Wrigley (1962) emphasised that there was a transition from an organic to an inorganic raw material basis, and in the first years of the 20th century Sombart spoke of a transition from a 'wooden' to an 'iron age'. With the utilisation of coal the bottleneck that had until then slowed down all technical economic innovations was overcome. Only then was the direction set for a process that economists today call 'economic growth', which rests upon natural conditions that cannot be reproduced historically and must necessarily be transitory.

IV

Germany in the 18th Century:
Wood Crisis and Strategies for Solutions

The breakthrough of coal use in England was the result of two factors: natural geographical preconditions and a technological and economic dynamic that progressed inexorably to the Industrial Revolution. The regional energy shortages caused by transportation difficulties, typical of agrarian society, led under exceptional circumstances to a shift to more intense use of fossil energy and the development of suitable production techniques. Nascent capitalist society and the emerging market made available an information and decision-making system that encouraged the most cost-efficient procedures based on coal. When there was a massive transition to coal in Britain, the market was well enough developed to further this process. At the same time there was no social or political institution that could have slowed down or prevented the process, as had been the case in China, where approaches to fossil energy were not followed up. Also, there were no institutions that could have sponsored this process in a directed way.

Long before there could have been talk of a desperate shortage of energy, the use of coal was widespread in England. The situation on the continent, for example in Germany, had a different pattern. First, geographical and natural preconditions must be cited. More land existed that was not well suited for uses other than forestry. Transport conditions favoured rafting wood and hindered the transportation of coal. But there were also grave social and political differences.

It has been noted that England established an internal system of checks and balances no later than 1688/89, resting upon a compromise of the ruling classes. Socially, the compromise was based on a rapprochement between the land-owning aristocracy and the commercial middle class. It did not insist on the sharp separation of aristocracy and commons that was customary on the continent. Wealthy lords were active in the commercial and financial operations of the city; merchants who had acquired wealth reached the goal of their life when they bought an estate and were accepted into the gentry. Social mobility through money was considered legitimate. The privileges of the aristocracy were comparatively minor. Attempts to establish an absolutist state were shipwrecked on the determined resistance of rebellious estates.

This was different on the continent, especially in Germany. In the territorial states of the German Empire a deep chasm separated the bourgeoisie involved in commerce and trade from the aristocracy, whose social position in the countryside largely remained unbroken. Particular classes were separated from each other by jealously guarded, complex regulations and privileges. The absolutist princes had a superior position. The aristocracy, which felt threatened by the rising bourgeoisie representing a new order of commerce and trade, found a guarantor of its socially superior position in the prince, even if it had to pay by being stripped in of its political powers. The bourgeoisie felt it was protected from aristocratic transgressions and the plebeian lower class by princely regulations. Within the rather broad frame that was allowed to them by these estates, the courts of absolutist princes could act as they pleased.

Economically, this meant that a special role was granted to economic and welfare policies in absolute states. In England the market became the gravitational centre of economic activity. In the end it decided if a particular production technique would establish itself or not. A comparable emancipation of the market from state regulation did not exist in German territories. It was the duty of the state's cameral and police authorities to order separate economic operations into a coherent whole. Care for the proper nourishment and welfare of the people remained a political function. It was not taken as a matter of course that the economic actions of individuals would combine into a harmonious whole by themselves. The principle of market integration had not established itself. If a crisis could be foreseen, it was expected that it was the duty of the state to find an answer.

If one considers that the advantages of coal in terms of energy were not as clear cut as in England, it is not astonishing that the transition to coal occurred later, was slower and was mediated by the state. Even if the market had played as central a role in Germany as in England, the substitution of wood by coal would not have taken place as early and as generally. If this process was characterised by an almost organic, slow and unplanned growth in England, in Germany it had the character of a deliberate and enforced act of violence by the state. It began later and was initiated by many administrative measures.

Despite the strong differences between England and the European continent there were shared elements. The continental states also participated in the global poker game of power expansion and even landlocked petty German states attempted to strengthen their individual positions against competitors in every way possible. They attempted by every means to support their respective domestic trades, to increase population numbers and to fill state coffers. This development policy appears to have hit problems of energy resources early in the 18th century. In some regions that would be felt as a wood crisis.

A series of documents indicates that the scarcity of wood increased in Germany over the 18th century, and threatened to become a general raw material crisis, no longer restricted to single regions and difficult transport conditions. The provision of wood was endangered from several sides. Population grew rapidly after the devastation of the Thirty Years War, not least because of the demographic policy of the state. Between 1700 and 1800 it rose from 15 million to 24 million. At the same time the per capita wood consumption remained high because people still acted as if wood was superabundant. Since commercial use also increased, absolute consumption rose tremendously. Everywhere signs of scarcity were observed, and contemporaries who attempted to explain them were not far from the truth:

> It is known even without my remembrance, how much all Germany, and especially worthy Saxony, has increased in great quantities of all things good due to a long enjoyed peace and sparing, for which we cannot thank the Most Highest and Giver enough. The deserted and abandoned villages were largely rebuilt, the arable cleaned, so that where before there was bush and hedge, now there is the most beautiful grain that can be grown. No less have the people multiplied, and they have become many, by whom the land is cultivated and ordered to provide all in abundance. Yet, despite this fortune, there is a small inconvenience because through cleaning and clearing the arable and meadows almost everywhere a scarcity of wood will arise, so that to supply it, a great deficiency is felt in these middling countries (Leutmann 1720, Vorrede).

While in all these areas the demand for wood grew, the condition of forests declined. As we have seen, the forest was part of the peasantry's economy. With the growth of the rural population, subsidiary use of the forest by peasants increased. Mercantilist economic policy also led to a heavy burden on the forests. In that doctrine, wood was only an auxiliary means to production. Its plantation could only indirectly contribute to increasing the wealth of the country. Mercantilist forest production therefore tended to keep wood prices down artificially, especially by fixing prices. Goods manufactured using wood – this was the case in almost all branches of the trades – were to be produced as inexpensively as possible so that better opportunities for sale abroad would exist.

Occasionally mercantilist policy violated its own rule not to sell raw materials abroad. If the coffers of the princes were once again particularly light, timber was exported to the Netherlands or England. Logs were rafted from the Black Forest or the Mark Brandenburg down the Rhine or the Elbe and sold to European naval powers.

In the 18th century many contemporaries looked woefully back to a past in which there had been wood in abundance and complained about the poor state of the forests. Scarcity of wood and woodlands that looked more like heaths were interpreted as signs of a poor economic policy:

> Felling of wood has increased to such an extent for some time that almost everywhere bare mountains and clear cut forests reveal to everyone their poverty of wood and accuse their inhabitants of bad housekeeping before the Creator (Zedler, vol. 52, 1747, 1162).

Considering the tendency to overuse forests, while at the same time wood production was encumbered by the subsidiary uses of the peasantry, which were fixed as privileges of common usufruct, a general scarcity of wood appeared to loom in the 18th century. The problem of a general wood crisis that was either acute in the present or threatened in the immediate future was reflected in a number of contemporary documents. Furthermore, there are considerations that support this supposition. According to one calculation (Mitscherlich 1963, 13), the annual yield of wood in Germany was only 1.5–2 m³/ha woodland because of the poor state of forests instead of the theoretically possible 4–8 m³/ha with optimal and exclusive usage of forests. Since only a third of the country was covered with forest, this amounted to 50–70 m³/km² total land surface. If a population of 44 persons per km² is assumed, only 1.5 m³ per capita were available annually. In the face of heavy commercial use this figure surely lay below the long-term need. If, in this light, the current consumption of wood was to be sustained, the time could be foreseen when the existing wood supply would be exhausted and limitation to the sustainable amount would become necessary.

Lack of wood was a general phenomenon on the continent in the 18th century. According to an estimate of 1811, Denmark annually consumed 1,720,000 *favne* (= 3,784,000 m³) firewood. That is just 1.75 m³ firewood per person, which was considered a sign of utmost scarcity (Kjaergaard 1994, 122). The Prussian forestry counsellor Friedrich August Ludwig von Burgsdorf presented in 1790 a treatise to the Prussian Academy of the Sciences that began with the following words:

> It appears utterly redundant to repeat proof that the consumption of wood is greater than the increase and increment with us (Burgsdorf 1790, 265).

Burgsdorf predicts that Prussia must anticipate the complete exhaustion of its wood resources in no more than 30 years unless reforestation with fast growing trees begins immediately. This scarcity of wood affected not only firewood but also timber, which must have a minimum age, while the demand for firewood could still be covered for some time at the expense of timber.

Numerous descriptions of the state of German woodlands from the second half of the 18th and the first half of the 19th century spoke of a frightening devastation and desolation of forests that extended even into the most remote areas. In the Palatinate, a cartographic survey done in the 1770s demonstrated that some areas that were officially designated forests barely had any trees. In some Prussian regions the oak forests had been cut to cultivate the land. The poor quality of the soil permitted no agricultural use of longer duration, so cultivation was abandoned again and the cleared land was left behind as waste and heath. Systematic reforestation was rare. Formerly forested areas became fallow where scrub and all sorts of bushes grew, but where trees could not prevail because of pasturing goats and sheep.

These desolations and the threatening lack of wood caused cameralistically schooled administrators of absolutist states to look for remedies. In the long run, the scarcity of wood also threatened to impede the interests of the state, i.e. expansion of power and taxation. Johann Heinrich Gottlob von Justi, who was a professor and economic advisor in Austrian, Hanoverian, Danish and Prussian service, directed the attention of cameralists to the forest as a wood producer:

> Woodlands are an important object to cameralists from two viewpoints. For one, they constitute a significant source of income to the state, and because in Northern countries they make an indispensable necessity for the preservation of their inhabitants without which such countries would be difficult to inhabit. (Justi 1761, 439f.).

Therefore, a particular part of the country needed to be exclusively dedicated to growing wood. It was important to determine the exact balance of arable and forests. On the one hand there should not be too many forests, since this would be at the expense of the arable and would prevent the maximum possible population numbers from being achieved. A population as large as possible was in the interest of cameralists, since that would increase the power of their state relative to others.

> Too few woodlands will also be a disadvantage for the population. ... For insufficient woodlands with a large population will have as a consequence that the price of wood will rise to excess. This has a negative effect on the price of goods from manufactures, factories and all other trades and their products, and this high price will be damaging to the export trade and, as a natural consequence, to the population. However, as soon as wood is at such high cost and therefore the most profitable use that one may have for land, then people will turn their efforts to wood plantation and the amount of the grain produced in the country will be reduced. (Justi 1761, 441).

Justi considers tree plantation, like Yarranton and Evelyn before him, as an alternative land use that competes with agriculture. But he does not draw the conclusion that the right proportion of arable and forest should be left to the market mechanism, though he sees a causal relationship in the substitution: scarcity of wood leads to a high price of wood that induces tree plantation at the expense of field cultivation. However, he wants to prevent this process by improving wood supply administratively. Apart from the fact that an unreserved faith in optimal allocation of factors of production by the market was not at that time widespread among economists, a decision for one or the other form of land use had rather long-term consequences, so that the need for planning by the state appeared evident. The same applied to the establishment of trades with a large demand for wood. Their profitability was not considered sufficient proof that they were beneficial to the common good as the representatives of the absolutist welfare state understood it:

> The Prince, due to his forest sovereignty, is entitled...to the right of direction in this matter; because only he and his cameral and police colleges have the insight and may judge whether the establishment of such works is useful or if these works give rise to damage due to their excessive multiplication or location at an inconvenient site (Krünitz, 43, 1788, 915f.).

Therefore, it seemed evident from the start that a governmental solution for the dreaded wood crisis had to be found. Most strategies for solution developed for this purpose emanated from government agencies. Only when the problems were reasonably well known to the public did private persons and private corporations begin to look for solutions independent of the state. In retrospect three strategies for a solution to the wood crisis, which were almost never independently promoted, may be distinguished (cf. Gleitsmann 1980): (1) Measures to conserve wood; (2) Functional separation of forestry and agriculture, that is development of rational forestry by monopolising woodlands for wood production; and (3) Substitution of wood, mainly by coal. These three solution strategies will be considered more closely.

1. Conserving Wood

As we have seen above, certain restrictions on the consumption of wood, mainly intended to spare mast-bearing trees, already existed in the Middle Ages. The multitude of decrees and measures to conserve wood that accumulated in the 18th century were aimed at all facets of wood consumption. The conservation catalogues, as they were set down in forest regulations but also in a series of other regulations and recommendations, addressed various forms of wood use without systematic differentiation. In retrospect, the demands may roughly be classified in two fields:

• regulations directed towards conservation through abstinence from consumption and

• regulations directed towards conservation through more efficient use.

Also, they could be distinguished with respect to whether they were aimed at wood as an energy source, as construction material or as chemical raw material. We shall not pursue these distinctions in detail in case the analysis becomes overly complex.

The conservation literature of the 18th century is full of complaints about the enormous use of wood by peasants. Thus, it is said in an anonymous writing with the title 'Of Germany's wood-wasting abuses, how they are to be prevented and the art of wood conservation is most readily achieved':

> How sad the wood-wasting chambers, baking ovens and fireplaces look in countryside and town. In some regions the rustic has tile stoves of horrible size. ... In the Vogtland and the Saxon Erzgebirge [Ore Mountains] chambers are heated in summer and winter, people cook, bake, heat water for the cattle, warm milk on the stove and doors and windows are commonly left open, despite some hammer mills already being affected by a scarcity of wood (cit. Ertle 1957, 1246).

The simplest form of regulating conservation was a state quota on wood consumption. It could only be enforced in towns since only there could the acquisition and consumption of wood be supervised. Special measures were enacted in Prussia. When the town of Berlin began to grow rapidly in the 17th century, complaints regarding a shortage of fuel, which was expressed in high wood prices, were soon voiced there. The state administration reacted to it by lowering wood prices with edicts but of course repeated lowering (1691 and 1693) did not result in the desired effect. Finally, a central firewood administration was established in 1694, which was responsible for the provision of wood from the demesne forests located near water routes to Berlin and especially to the authorities. The private wood trade was strictly regulated by law.

Königsberg went a step further. Here a firewood regulation of September 21, 1702 stipulated how high the consumption of firewood in a household could be 'according to its rank' and 'even the most essential trades such as bakers, brewers, cooks, butchers, potters etc. were given a particular fixed quota which they could not exceed under a heavy penalty and were consequently forced to cease their operations once it had been used' (Pfeil 1839, 111).

In the countryside a multitude of traditional forms of wasteful wood uses were forbidden. The catalogue of such orders is extensive. Thus, the forest order of the Electorate Hanover of 1768 ordered:

> The abuse that for a wedding or festivity, special trees are demanded for benches for people to sit on or as firewood shall be abolished and in each parish several benches shall be made, stored, loaned out and used at weddings and festivities, and then taken back to their assigned places and stored again (cit. Ertle 1957, 1245).

The attention of forest regulations was particularly directed at construction. Peasants who were erecting a new building were to make proposals that had to be examined for their appropriateness. Building was to be restricted to their basic needs and, in particular building for 'private amusement' was forbidden (Schwappach 1886, 360). The police authorities were to conduct regular building inspections to determine defects and make provisions for their rectification, since it was feared that due to neglect more repairs would become necessary and consume more wood. Other regulations determined that living hedges were to be planted or ditches dug in the place of board fences. One author proposed to reduce the size of rooms so that they would be easier to heat:

> What unavoidable necessity urges us to choose large and high rooms as living quarters in winter. These should be done away with and such should be chosen that are small and low, then one will clearly…. save much wood. … I do not speak of the palaces of great lords here. Small low rooms would not be appropriate for them. Nor do they need warmer rooms. Many strongly spiced dishes and spirited drinks that they enjoy bring so many sulphurous parts into their blood that warm rooms are unbearable to them anyway (Möller 1750, 667).

Only common households in the towns and peasants were addressed. The latter found an advocate in the conservative magistrate and writer Justus Möser, who spoke up 'for the warm chambers of countryfolk'. But, defenders of civic domestic culture also rejected such curtailing of comfort. Thus, the propagandists of energy conservation, who promoted small and low rooms, faced accusations:

> They seem to approve of the old lifestyle when in most houses husband, wife, children, servants, maids and even calves and piglets all lived together in one room (Krünitz 43, 1788, 922).

This comment makes evident that the measures and suggestions for energy conservation met resistance and there were voices that turned against a too strong regimentation of domestic conditions. The absolutist state tended in any case to meddle in the private sphere of the people and complaints about scarcity of wood could be viewed as an excuse for extending the sphere of political regulation at the expense of the citizen's freedom. Georg Friedrich Möller, for example, whose suggestion it was to reduce rooms for the sake of saving wood, assumed as a matter of course 'that such suggestions cannot be enforced unless the police takes the matter into its hands' (Möller 1750, 683). However, it is well to bear in mind that *Polizey* meant the branch of administration in charge of public order, not the police in our meaning of the word. In particular, more liberal-minded economic writers turned against the regimentation madness and pointed out that extreme measures would have a diametrically opposed effect to the one desired.

Thus it would mean to insult the natural freedom of man if he were subjected
to force here; not to speak of finding a suitable means to make such force reality
(Krünitz 43, 1788, 922f.).

More moderate suggestions tended towards regulation of wood conservation if this could be easily supervised. Thus, Krünitz advised to permit
building of only three-storey houses in towns because 'to roof a three storey
house required no more rafters than a one storey house' (ibid., 924). Furthermore, it was recommended not to shape boards, beams and troughs with an adze
anymore, but to saw and glue them because that would result in less waste. Also,
there was no lack of oddities: in Austria it was commanded under Joseph II that
the dead should not be buried in coffins but in black cloth (Schwappach 1886,
362). Georg Christoph Lichtenberg, too, had an original suggestion:

> The forests are shrinking, wood becomes scarce, what shall we do? Oh, when the
> forests will finally have come to an end, we surely can burn books until new ones
> have grown (Lichtenberg 1776, 495).

In any case, state organs attempted to enforce energy conservation.
There are lots of sources with recommendations or even regulations for
employing the newest 'arts of wood conservation', especially in domestic
heating. First, emphasis was placed on better insulation of rooms and houses
since this would immediately achieve a conserving effect without lowering the
quality of life. Thus, it was said in a higher authority commission (Ober-Amt-
Patent) of the Margraveship Upper Lusatia of August 20, 1767:

> Poorly kept chambers, especially among subjects in the country, cause futile
> wood consumption. Therefore, the chambers shall be kept with covering of good
> panelling and in good repair from the outside and during the customary and
> where possible biannual inspection of firewalls and stoves it shall be eagerly
> sought if... this regulation is obeyed (Schmid 1839, 183).

The intention was not merely that all houses were to be insulated against heat
loss, but that the condition of the insulation should be inspected regularly.
However, such regulations and general appeals 'to conserve and save firewood'
(Schmid 1839, 182) were found in all forest orders of the 18th century –
probably an indicator that they were not well obeyed. For peasants, who had
traditional rights to take firewood from communal woods, it was not apparent
why they should do without something they had always been entitled to. Even
regulations that only dry wood blown down by the wind was to be collected as
fuel were obeyed only reluctantly, under threat of severe penalties.

Fuel conserving stoves

The invention of a multitude of conserving stoves constituted a focus of efforts to conserve through more efficient energy use. We have mentioned before that the origin of domestic heating may be sought in an open hearth without separate smoke draught from which the open chimney and the iron plate stove developed in due time. All these stove types were very inefficient in terms of energy: the bulk of heat escaped outside when heating with an open hearth. The problem in the construction of conserving stoves consisted in providing sufficient draught for smoke while simultaneously keeping the warm air inside the room. The stove had to be built in such a way that as much heat as possible radiated into the room but preserved a sufficient temperature gradient to carry smoke outside. Emmerich Kulmann in Passau built the first known conserving stove in 1325. Patents followed in the years 1557 and 1562.

One of the oldest propaganda pamphlets on conserving stoves was the 'Wood-Conserving Art' (1618) of Franz Kessler of Frankfurt-am-Main. The Thirty Years War interrupted this search for fuel-conserving stoves. Only in the 18th century did the problem become so acute that a whole flood of writings appeared containing projects for the construction of wood-conserving stoves. In 1763 an official contest was conducted in Prussia for 'a stove for rooms that consumes little wood' and in Berlin a 'Society of the Wood-Conserving Art' existed in 1784–87, which published a series of writings 'Of the Use of Wood-Conserving Stoves' (*Allgemeine Deutsche Bibliothek* 83, 1788, 2, 583).

Various types of conserving stoves were, as a rule, constructed by a purely empirical route and their building instructions could often not be applied properly. Too much depended on the immediate circumstances under which they were built and too little was capable of generalisation. According to Gleitsmann 'the theoretical value of the conserving stove literature should be rated higher than its pragmatic value' (Gleitsmann 1980, 124), i.e. the effect of the conserving stove propaganda on the public had a more important effect than its actual distribution. The fault criticised by a review on the series of the Berlin 'Society for the Wood Conserving Art' probably applied to other representatives of the wood-conserving arts as well:

> The practical part of these pamphlets has been such that one has no cause…to doubt the benefit they occasioned. All the more reason to wish that the whole was more planned, the style more correct and clear, the theories of the authors were more informed by physics, especially with respect to the nature of fire and air, by which the good philanthropic purpose were effected more easily (*Allgemeine Deutsche Bibliothek* 83, 1788, 2, 583).

When these stoves were actually built, the conserving effect was on occasion significant. Thus, in the late 18th century a conserving chimney reduced wood consumption to a little more than a third of ordinary consumption. Another obstacle to the wider distribution of these conserving stoves – apart from technical difficulties in reproducing them – was their price. Balthasar Heinrich Baron de Heins noted this point in a letter to the Royal Prussian General Directorate in 1764:

> Thus, the question who profits from such wood conserving stoves and can make them useful? The answer is none other than who has the money to have such stoves put in for 20, 30, 40 pp. Reichsthaler (cit. Gleitsmann 1980, 124).

The authors of pamphlets promoting wood-conserving stoves often spoke of craftsmen passively resisting during their construction or not being able to understand the building instructions. In particular, Leutmann (1720) returns again and again to the topic of how difficult it was to get craftsmen to build something that was outside their normal experience. This was probably a fundamental problem for inventors of conserving technologies: on the one hand they were directing their books at an educated, moneyed public that could not only afford to buy a book on conserving stoves but was also open-minded enough to invest in such an enterprise. This educated public expected the author to be at the apex of the physical knowledge of the day and to derive the wood-conserving stove from the 'nature of fire'. The conserving pamphlets were in part supplied with extensive building instructions that were added to the book as a special appendix. These instructions were primarily directed at craftsmen, who were mostly empirically trained and had no idea of the physical theories addressed. If the conservation-minded contractor did not understand himself how to interpret the instructions, effect the implementation of the plan in person and see to its correct execution, there was little chance of the conserving stove ever functioning.

The need for special operation and maintenance often arose with particularly expensive and complicated conserving stoves. In households that could afford conserving stoves, heating was a duty of servants, who were often not able or willing to follow the operating instructions. There were many complaints that conserving stoves had no proper draught and that they were smoky and sooty. It was usually the servants, who viewed these newfangled devices with distrust, who were blamed.

Of course, there was considerable charlatanry with this new wave of conserving stoves from the mid-18th century. The most varied inventors and projectors presented plans for stoves that never worked and could not work. Apparently, a kind of fad market for conserving stoves had developed that was less interested in economic utility than in technical novelty. I know of no contemporary attempt to test the returns on a conserving stove. If one looks at building instructions and illustrations of completed stoves (see Faber 1940a; 1940b; 1941; 1957), one comes away with the impression that conservation must have played a subordinate role in their construction. The effort alone for all sorts of decorating and painting must have been so substantial that savings in the cost of firewood must have been more than compensated. In any case, the highly complicated spiral stoves were out of bounds for the households of the less wealthy.

Towards the end of the 18th century, attempts to develop simpler saving methods that could be used by the broad population increased in number. Thus, C.H. Meisner wrote a 'Manual of Wood Conservation' in 1801 with the express goal 'of affecting something good…for that class of person that is most burdened by the increased expense of wood' (Introduction). This and other writings that appealed to a broader audience were different from the bulk of wood-conserving literature because the volumes were slimmer and cheaper and usually contained no elaborate illustrations with engraved building instructions. In the first place they counselled conservation through abstinence in wood consumption but there were also suggestions for technical improvement of stoves that could be carried out with little expense. However, these pamphlets emphasised promoting the use of mineral coal as a wood substitute. We will return to this problem.

State administrations observed the development of conserving stoves with interest. The drafts of a new chief forest regulation in Carinthia in 1755 contained, for example, a detailed description of a stove type that rural households should use to save wood. This stove was supposed to be narrower and higher than older models. Furthermore, it was intended that it should stand freely in the room, not in a corner, so that the heat could radiate unobstructed into all directions. The design envisioned inspection of the stoves by the state. Where stoves of the older construction style were found, they were to be smashed and their owners fined. However, this far-reaching design was never put into practice.

Problems in the commercial field

In the commercial field, too, attempts were made to limit wood consumption in response to rising prices. In Berlin the price for a *Haufen* of pine wood (c. 15 m³) had risen six fold between 1693 and 1767. With wood prices increasing one would assume that no special regulatory measures were required to motivate trades and manufactures to be more economical with wood. However, what happened was this: export-oriented and strategically important branches of industry were granted wood at an especially low price. In Carinthia the owners of woods that were located near iron works were forced to provide them with wood at a low, fixed price. Other uses of the forests, in particular clearances, were prohibited. In some regions the wood or charcoal yield of a forest was expressly reserved for a particular iron works. This 'forest or charcoal dedication' meant that the privileged consumer could dictate the price. It was a death sentence for the local iron trades in Carinthia when the forest dedication was lifted and the wood market freed in 1783 (Johann 1968, 196). As soon as they could no longer fix the price, the iron works had to compete with other wood buyers in the market. As long as the state kept the price for (energy) raw materials artificially low in the spirit of mercantilist economic policy, iron works were able to export. When freedom of trade and commerce were established, they were suddenly exposed to market forces, which finally ruined them.

This was exactly the consequence that German cameralists had always feared: if the market were liberated, domestic industry would be destroyed and those employed would lose their livelihood. That was the reason why economist politicians turned against proposals that energy consumption could regulate itself through free pricing. G.F. Möller had already argued in the spirit of such a market economy around 1750 that it was

> ...under present circumstance, when so much depended on the conservation of wood, better to set the price somewhat higher rather than lower. Nothing serves better to spur people to think of all sorts of means to do with less than necessity. (Möller 1750, 681f.).

Against socio-political objections that an increase in wood prices would particularly affect the poor and was therefore inhumane, he answered that 'philanthropy must not merely be extended to those present but also to those absent' (ibid.). Fuel conservation through charging the present generation will be advantageous to future generations.

Cameralist welfare and economic politicians had grave concerns with respect to this position. They would not have shied away from requiring the poor to abstain from consumption – but reduction in wood consumption for manufacturing, and thereby of (exportable) production, should be avoided. That the trades would react with technical innovations when raw material prices increase – a thought that appears obvious today – was explicitly rejected:

> To expect of producers and manufacturers the invention of useful wood saving arts must be a futile hope because this would require men who understood the mechanism of fire. But if such did invent useful wood saving arts and made them known, then those, who think at all times of their advantage, will accept them on their own without the police forcing them to this with a higher price of wood. (Krünitz 43, 1788, 923f.).

The argument was as follows: private households may be brought to limit their consumption when wood prices are raised but it must be anticipated that they will not achieve this reduction by more efficient use but by denying themselves a particular benefit (own coffin, warm home). Commercial consumers must be expected to behave similarly with the consequence that they would restrict production or that their production costs would rise, which would diminish their export chances. Both should be avoided.

Therefore, export firms were distinguished from those that merely produced for domestic demand. High wood consumption for brick works, potteries, bakeries and distilleries was considered pure waste and consumption restrictions could be applied to them. Export-oriented firms were to be spared. Today it would be expected that raising the price of firewood would provide an incentive to develop methods of conserving fuel. But none of the authors I am familiar with assumed this. They thought manufacturers would accept technical innovations if they were offered but not that they would develop them of their own accord. Inventions were the reserve of specialists, of tinkerers and project makers, who were supported directly by the state. Since traditional crafts or manufactures that were not oriented to growth or to market conditions did not possess the means to experiment, it was suggested that the state should at least assume the *costs* for this:

> Every citizen shies from them [the costs] especially if they are high. No one wishes to attempt experiments for fear that they do not go well because then money, time, effort and labour are lost. If the Police wish to achieve their purpose, they must effect that experiments are undertaken in princely or municipal breweries and distilleries... When citizens and private persons, who have their own breweries or distilleries, see and are assured by received news that such a new installation will save much wood, they will join on their own accord, because it is to their profit which is their motivation for undertaking such a business. The police would not even need to use coercive means...(Krünitz, 43, 1788, 926).

The rather liberal-minded economist Krünitz trusted that within a suitable governmental framework free pursuit of economic self-interest would cause individuals to adopt conservation techniques sponsored by the authorities. He turned not only to the trades, who were to be convinced that conserving wood was in their interest, but also directed his argument at magistrates who should be discouraged from regulating the economy too meticulously. Contrary to this procedure, in contemporary forest regulations conservation was often ordered in an authoritarian tone:

> The brewing stoves, malt kilns, bleaching stoves must be set up, if possible, according to the new inventions of the wood conserving art and painstaking care must be taken to achieve the resulting advantages of less expense and sparing wood to one's own advantage (Higher Authority Commission of the Margraveship Upper Lusatia, 20.8.1767, § 14, in: Schmid 1839, 184).

It seems somewhat comical to the modern reader when the state orders individuals that 'painstaking care must be taken to achieve...one's own advantage'. It appeared evident to economists influenced by physiocratic teachings that enlightened pursuit of economic self-interest would clearly result in a meaningful whole that could be described objectively, almost scientifically. But cameralist paternalism did not spring from deficient insight into the laws of economics. Rather, it may be seen as an indication that a rationally calculating position was not yet widespread enough among individuals to leave the creation of a functioning economic balance to the free play of economic forces. The opinion that the market was capable of creating a meaningfully ordered whole was present only in its infancy. Economic integration of society was not yet entirely mediated through objective market mechanisms, and this was reflected in economic policy, by requiring that the economic whole had to be established beside individual self-interest as a separate entity. The rationality of the economy was not constituted in a self-structuring sphere of objective mechanisms but had to be put into force as an autonomous goal.

It is difficult to decide between what is simply the ideology of power, the mere wish to extend the ruler's authority, and what stemmed from the structural condition of an economy that was not yet differentiated as a self-regulating social subsystem. We encountered a similar problem with respect to the promulgation of forest regulations. We are at the historical transition between normative and market integrations of society. The elements of normative regulation are recognisable because the state as a welfare and police state understood its role in a normative sense, but confronted the individual with these norms entirely externally, using laws and regulations and therefore indirectly accelerated the differentiation of structurally autonomous fields.

The mechanism of market integration, then, was only rudimentarily present, with the state repeatedly drawing legitimisation from these deficiencies. In particular in the energy and raw material crisis it is apparent that German society was not capable of providing a spontaneous answer in terms of market integration, which was possible in England (albeit under exceptional natural conditions). Only when the transition to fossil fuels had taken place on a large scale did regulatory details become obsolete. A state energy policy that enforced conservation with coercive measures and supervised it remained an episode. However, from the contemporary perspective it was not apparent that these were the alternatives of the future: a growth oriented market economy based on fossil energy or a solar energy system that mediated conservation by a multiplicity of means on its way to a stationary sustainable state.

A new source of energy may have been opened up through energy conservation as a response to the looming wood crisis of the 18th century, but this was a source that would quickly run dry by its very nature. Rooms could be insulated and stove efficiency improved. Wood-conserving arts of all kinds could be employed in trade operations and it was possible to reduce wood consumption in non-energy uses so that more firewood was available. Still, all these endeavours had natural limits. Nowhere is the law of diminishing marginal returns more in effect than in energy conservation within the traditional framework of the solar energy system. All conservation measures could only have delayed the time at which the energy potential of a particular area was fully exploited. At some point the stationary upper limit would have been encountered and further energetic expansion would have been precluded. This does not mean that the energy system would have collapsed. A solar energy system cannot collapse as long as the sun shines. But the commercial dynamic that culminated in the Industrial Revolution could have been throttled by energy scarcity. Instead of the sensational growth of the last 200 years there would have been increased struggles over the distribution of scarce resources, in which assigning bureaucracies would have played a role that would essentially not have been different from that of an absolutist welfare state.

2. Functional Separation of Agriculture and Forestry

Another strategy in the looming wood crisis was to increase the wood supply. We have seen that in 18th century Germany the real annual increase of wood per hectare of land amounted to c. 1.5–3 m^3, while the biological potential of wood growth was at least 5 m^3. Various supplementary agricultural uses of the forest were the reason for this difference. The forest was not in our sense an exclusive forestry zone reserved for controlled growth of trees but formed an element of agrarian economic space within which different forms of use could not be precisely isolated. If it was possible to uncouple agriculture from forests without changing proportions, wood yield could be increased up to threefold.

The process of functionally separating agriculture and forestry rested upon three elements that were relatively independent of each other. The most important were:

- Termination of peasant rights of common use (privileges of usufruct);

- Development of new methods of agricultural production that did not require use of the woodlands;

- Development of rational forestry (forest sciences).

The onset of the functional separation of agriculture and forestry cannot be exactly dated. Origins may be found in forest regulations of the 16th century, which attempted to limit supplementary agricultural uses, and in earlier attempts at encouraging the growth of trees with systematic spacing of cuts and reseeding. Thus, the Elector Johann of Saxony's wood regulation of 1527 prohibited pasture in forests after clear-cutting until the young trees had grown two ells high. The Bavarian forest and hunt regulation of 1568 prohibited unrestrained goat pasture in the forest. Individual peasants kept as many as 100 goats that were taken to pasture in the forest. Attempts to limit their number met heavy resistance because they endangered peasant subsistence (Johann 1968, 38). The forest regulation of Brunswick-Lunenburg of 1680 envisioned protecting young nurseries until they had outgrown livestock browsing. According to regulations of Baden in 1712 and 1748, goat pasture was to be permitted to the poor and widows, but only in special designated places (Bernhardt 1872, 233). Similar prohibitions and restrictions are found in many forestry orders and are often repeated, which indicates that they were rarely obeyed. The fact that the game stock was very high in forests gave peasants little reason to relinquish their ancient herding right. The conflict between hunters and poachers, between foresters and forest violators is a rock-solid substrate in traditional European societies.

Forestry regulations were in part enforced with the threat of heavy penalties. Forced labour, incarceration, even the loss of the right hand or the death penalty were imposed against severe trespasses, especially poaching but also arson in the forest and felling of trees (Bernhardt 1872, 235). Their enforcement met considerable resistance that could turn into local uprisings. The rural population merely considered transgressions against the princely forest as justified by custom, without any sense of wrongdoing. This was demonstrated by wood theft legislation in particular. Thus, the Prussian law of 1821 wanted to permit forest owners 'to restrict wood collection to the destitute or to prohibit it altogether'. The consequence of this law was a sudden leap in crime rates since an action was now outlawed that had been traditionally considered as permitted. Contemporaries noted 'the widespread opinion among people that wood grows for all and that taking wood from the forest was no sin' (cit. Blasius 1975, 110 f.). Successful poachers and wood thieves were often celebrated folk heroes, whose lives were surrounded by legends reminiscent of the tales of Robin Hood. The myth of the noble and fearless poacher was an escape valve through which the rebellious imagination of rustics vented its displeasure against princely oppression. At the same time the guerrilla warfare between forestry officials and poachers was a buffering mechanism of social protest, a social rebellion without consequence.

To the peasant the forest as a component of agricultural life and production space was part of nature. Wood was a vital natural product of nature that could no more be absent from the rural economy than water, sun and air. It did not have to be produced through hard labour but was present 'by nature'.

It was a last remnant of the life as hunter-gatherers. Hunting was now forbidden by royal edict and the population was too dense to be sustained by fruit-gathering. Fruits that were once collected free and without investment were now wrested from the soil by hard work. Cattle that had been hunted by free men were now tended and pastured. Only wood (and maybe berries, herbs and mushrooms) could be gathered at will. Thus, wood scarcity with its many prohibitions and restrictions on rights of use led to the transformation of the forest into a preserve of forestry – a zone of control and work – and thus completed the transition to agriculture. In world historical terms the shift to controlled forestry finally completed the Neolithic Revolution. The agrarian solar energy regime became totally controlled – at a time when the transformation of agrarian production itself was emerging.

After the closing of the forest only water, air and sun remained free goods provided by nature. Woods had become products of labour, which now belonged not to nature, but to society and the economy. No wonder if peasants opposed this development. It is no accident that the following prediction was ascribed to Luther, that the Last Day was coming when three things were seen to be lacking: good friends, good coin and good wood. When even wood was scarce, the end of the world must be near.

The most important agricultural use of woods was for pig forage, which still predominated in many remote areas. In Brandenburg a proclamation prescribed that the peasants had to pay a rent (pannage) for every pig in the Elector's forest. This pannage was the chief income derived from old oak forests (Pfeil 1839, 68). Landlords and peasants had a common interest in the pasturing of pigs, which was conducted at the expense of wood growth. Peasants could only stop driving pigs into the forest when alternative fodder for the animals became available. This was the case at the close of the 18th century when new types of fodder plants were grown (potatoes, fodder beets, clover for cattle), a process that only established itself in the course of the 19th century.

A transition to stall-feeding provided more dung but at the same time required more bedding for the stalls. For this purpose dry leaves were taken from the forest to put down and afterwards taken to the fields for manure. This practice of raking bedding removed valuable minerals from the woods, causing forest officials to attempt to restrict it, though they would not suppress it. Cutting grass, too, was increasingly forbidden. In the opinion of foresters, the many rights of use or servitudes that lay upon the forest were a serious impediment to rational forestry:

> Who does not know since the wish stirred to put woodlands into a better state that it was the servitudes or land rights attached to it alone that made its implementation impossible? Like an army of vultures they gnaw at the marrow

of the forests. Raking bedding, wood rights, herding, the right to keep wild bees, whatever their names, they truly contribute to turning the most beautiful forests into deserts and prevent that deserts are turned into forests. (Pfeil 1816, 76).

The termination of servitudes was only completed in the first half of the 19th century during the division of the commons and termination of burdens on peasants. Only the extensive privatisation of land use that took place in the course of the 19th century opened the route to a complete separation of agriculture and forestry. Efforts in this direction were made much earlier, but came up against the problem that a separation of land use had to be linked to an extensive restructuring of the rural economy.

Parallel to the process which separated the forest from agricultural production went the development of scientific and economic methods in forestry. A number of empirical insights on growth conditions in woods were incorporated even into older forest regulations but they were unsystematic and often wrong. In the mid 18th century forestry science experienced a first flowering. A multitude of writings appeared and a series of experiments was attempted to systematise knowledge. An essential prerequisite for the desired sustainability was an exact survey of the stocks and an estimation of gains and yields. Wood mass calculations and forestry mathematics were the first branches of scientific forestry. One result of a more exact calculation could be proof that the wood crisis was not as acute as it appeared in the first place. Apart from a calculation of the wood stock and its increase, the demand for wood had to be estimated. Justi had already pointed this out forcibly in 1761:

> In the first place it is necessary that one reliably knows the consumption of wood in the country. ... Therefore, it is necessary that registers and reports from an inspection of all woods in the country, the prince's and those belonging to towns, communities and private persons, is turned in noting their size, constitution and the kinds of woods growing therein, in which is primarily determined how much wood can be felled yearly sustainably, economically and without ruin. (Justi 1761, 443f.).

Such a survey of all woods according to increase and stock was difficult and inexact given forestry methods of the time. Even surveying consumption required a huge effort. Still such attempts were undertaken. Thus, a complete register of the inhabitants, who were asked what their annual demand in wood was, was compiled for the wood demand survey of the town of Munich in 1791. It ranged from 2–5 cords in poor households through 40 cords in middle class houses to several hundred cords with a maximum of 838 cords in aristocratic households. 65,000 cords were calculated as the total demand, with a cord equalling to around 3 m³. This quantity was particularly worrying because of the transport problem. The opening of new forests was therefore, begun in 1795 by building wood paths. Then a survey of forests finally showed that sufficient woods were available: The spectre of wood scarcity vanished with the dawn of a new age. Calculation quickly demonstrated that an abundance of wood was available where scarcity had once loomed. A contemporary commented on this state of affairs: 'It was feared one would freeze in the midst of forests' (Köstler 1934, 107).

However, wood scarcity did not turn out everywhere, as in Munich, to be an imagined phenomenon. In all regions forms of forestry were developed which approached the biologically possible growth rate of wood. Experiments were undertaken with establishing cutting sections; people became interested in forest biology, tried out exotic trees. These new methods were not always successful. Thus, large monocultures were planted but nothing could be done against massive attacks of pests. Calculations of future wood demand also went wrong on occasion. Early in the 19th century it was assumed that the need for firewood especially would rise, but the spread of fossil fuel resulted in an entirely different scenario. Beech forests that had been planted did not yield their expected profits given the decline in firewood prices.

The functional separation of the forest economy that turned it into a production system for wood or a field with a longer turnover period finally changed its character as a living space. That highly stocked, monotonous forest, as we know it, was created:

> In the forest, which once reverberated with the baying of hounds and the sounding horns of aristocratic hunts, the shouts of herdsmen, the bleating, neighing, bleating and grunting of livestock, the axes of wheelwrights and pole cutters, the pounding of hammer mills, where charcoal kilns, tar pots, ash pits and smelters smoked everywhere, became more and more silent. It was no longer living space but the site of planned and systematic production of timber intended only to provide as much and as valuable wood as possible. Planned forestry had arrived (Mitscherlich 1963, 15).

In summary, it can be said that in response to the wood crisis the attempt was not to limit consumption of wood by doing without or by a more efficient use, but to increase the actual amount of wood. This process was closely associated with the establishment of governmental forest regulations and the end of traditional rights, which was another reason why smallholding was deprived of its basis. The formation of scientific forestry can be interpreted as the attempt to dissociate wood production from peasant farming, which occurred at the expense of traditional agriculture and caused processes of adjustment there.

3. Substitution for Wood

The first two reactions to the wood crisis of the 18th century – conservation and rational forestry – were not systematically differentiated from the third, substitution of wood by other raw materials. Still, a process was initiated here that led to a completely new energy system. Substituting wood with coal not only used the current energy intake of the biosphere, but also accessed the stored energy reserves that had formed over millions of years. An age of relative energy superabundance began that permitted an increase of wealth and made it possible to feed a population explosion beyond belief. We have seen that this epochal process was initiated in England. On the continent the English way was imitated after a delay of 200 years. Germans came to use mineral coal in larger quantities in a frenzied attempt to find a way out of an acute or imminent shortage of wood.

The substitution of wood by coal was for contemporaries a process that was in principle no different from other substitution processes. It was not only in its function as a fuel that wood was replaced; materials were also sought to take its place in other uses. As a construction material wood could in theory have been replaced by iron, but we have seen that 50 cubic metres of wood were required to make a ton of wrought iron. If, therefore, a wooden beam was displaced by an iron support, it was not associated with any savings of wood. It is intuitively obvious that the charcoal obtainable from a single beam (or from the coppice that would have taken the place of the high forest from which the beam came) could never have smelted enough iron to make a beam. Iron was therefore only employed when a particular task could not be fulfilled by wood. Thus, in the early days of firepower there were wooden cannons, but it soon became apparent that it was worth the effort to make them from iron. Nevertheless, wood was entirely satisfactory for many other purposes.

Efforts to replace wood as a construction material were directed in the first place at building construction. Great savings were achieved when houses were made of stone. Of course, this was only possible if suitable quarries were available in the immediate area. The effort of transporting stones were far greater than wood since stones could not be rafted. As a rule it was cheaper to build with wood, especially if only the frame was made of beams as with half-timbered houses, whose spaces were filled with loam. On the other hand a stone house had a much higher life span so the effort paid over the long run. Less wealthy builders were not able to pay the expenses for a stone house even if it may have been less expensive than a wood house as a long-term investment.

Forestry regulations frequently prescribed that stones should be used to save wood, at least for those parts of the building that were particularly stressed. Thus, the Upper Lusatian Higher Authority Commission of 1767 ordered that it

…was to be seen to that the lower floor of residential buildings, but also of barns and stables, should be constructed, where possible, of stones and the upper floor be set or covered or at least faced with unburned bricks (Schmid 1839, 182).

The reference to unburned bricks illustrates a problem in wood substitution: if burnt bricks or tiles were used, the use of wood in firing was higher than the savings. In the short term it was not profitable to make brick houses. Even in England, building brick houses, which are often still characteristic of townscapes, only established itself in the 17th century, when it became possible to fire bricks with mineral coal. In Germany making bricks depended for some time on the use of firewood so brick houses were only used in areas were it was not possible to quarry stones nearby. There was only willingness to make the effort associated with bricks for special purposes, such as military fortifications, churches and publicly symbolic mercantile buildings. It is difficult to say if the use of bricks was more energy efficient than that of wood. This was probably the case with roofing since wood shingles only had a short life span but may not have been the case in the construction of upper floors that were not as exposed to dampness. We know how long oak beams will keep if they remain dry. Also, it was not possible to do without wooden beams and roof joists. The durability of a house was still determined by its wooden components but in this case it should be noted that when a house was torn down, beams were reused as well as bricks and tiles.

For certain house parts subject to considerable wear and tear, such as foundations and doorsteps, the use of stones paid in any case. This also applied to the making of well troughs and pipes, which were no longer made of hollow tree trunks but from fired clay. Another means of substitution was through living fences. Thus, Krünitz says

> That in Germany the usually customary dead hedges and dry fences, which are made of prunings, poles and planks, may be counted among avoidable waste of wood (Krünitz 43, 1788, 921).

And a Saxon General Order of August 2, 1763 prescribed:

> Plank and picket fences are not to be used for gardens and field separations but either ditches or lighter separations of dry poles, boards and waste are to be made, also living fences of blackthorn and hawthorn, hazels, beech, willows and such wood (Schmid 1839, 162).

As can be seen, no opportunity was missed to conserve wood, even if it was only to replace dead wood with living wood. Also, in using wood as a chemical raw material, substitutes were sought as far back as the 18th century and to some extent were even found. Firewood was required to generate heat in making soda from common salt but it was only a fraction of what was needed for the production of potash from wood. In this way the tremendous wood consumption of glassworks could be diminished without decreasing production. Here we have a form of technical progress that occurred independently of the shift to coal and brought considerable wood saving.

Substitution of wood by coal

The historically decisive escape from the wood crisis of the 18th century was the substitution of wood by coal. In the end this process resulted in such an enormous breakthrough in energy supply that other attempts to substitute and conserve appeared marginal by comparison. But for contemporaries it was only one way out among several – they were unaware of its epoch-making importance.

When the existence and possible use of mineral coal began to be noted in Germany in publications at the end of the 17th century and in the 18th century, direction was sought abroad and above all to England. The circumstances there were seen as proof that the technical problems of using coal could in principle be solved. The widespread use of coal in England was almost legendary. It was reported that

> …mineral coal was so prized there because of a lack of wood that beggars would rather take a piece of coal than bread (Bünting 1693, unpag. appendix).

Just how unknown coal was on the continent is evident by the fact that legends about its properties persisted for so long. Thus, the alchemist manual of Levinus Lemnius *De miraculis occultis naturae* stated in the late 17th century that a mineral coal fire could be ignited with water and extinguished with oil (Lemnius 1666, lib. I, cap. 17, p. 155). *Steen-colen* were described as a curiosity and it is apparent from the description that Lemnius never laid eyes upon them. Astonishingly, this legend can even be found in the otherwise strongly empirical work of Georg Agricola (1612, 473): 'Si aqua aspergitur, magis ardet: sin oleo restinguitur'. Even this expert in mining and smelting looked upon coal as a kind of mineralogical unicorn.

In this light it may be surprising that the origins of coal mining extend as far back in Germany as in England, if not further. Coal was already mined in the Aachen coalfield near Eschweiler and Weißweiler in the Roman period. Coal mining is documented in the year 1183 for the Wurm area near Aachen and in 1198 near Liège. First mentions of coal mining in the Ruhr district come from 1296 (Dortmund), 1317 (Essen), 1361 and 1375 (Duisburg). As in England, surfacing pockets were mined in the open. The start of mining proper dates from the 16th century and was of marginal importance. Coal production was a subsidiary peasant activity, especially in winter. Coals were used for one's own needs, in particular for heating (Aachen 1333) but also for certain crafts, such as smithing (documented in Dortmund 1389), lime burning, alum production and salt making. However, the use of coal was limited to the local scene until the 18th century. Thus, it is said in a mining treatise of 1768:

> One searches in vain for news of mineral coal in old mining books; for it has not been in Germany known for much time. (Scheidt 1768, 161).

Apart from the area around Aachen and the County Mark (Ruhr district), coal was dug in Saxony (from the 14th century), Silesia (since 1366) and the Saar district (first mention 1429). However, transportation considerably restricted the spread of coal use. Thus, the mining district of Mark supplied the blade works of Solingen and the vitriol and alum works in Schwelm. But no great distances could be covered on land if transport was to be worthwhile. The mining district had to be near a commercial buyer. In principle, there were more or less the same restrictions as for supply with wood or charcoal. Even to *Kohler*, who dug and sold mineral coal, no great difference between mineral coal and charcoal existed. Often they traded in both (Kürten 1972, 47).

In contrast to England, where deposits near Newcastle were found right on the Tyne and coal could easily be shipped by sea, an unfavourable transportation situation proved to be so restrictive in Germany that coal production experienced no significant increase until the 18th century. For a long time only opencast mining was practised and carried out without any expertise. Thus, it was said in a report from the Ruhr district of 1734/35:

> If it should please someone to produce coal, he will rarely join with fellows into an association, as is otherwise customary in well ordered mines, but will take an all too extensive territory as a claim but wager too little. Or, an individual peasant is struck by the fancy of gaining something thinking he can achieve his purpose the soonest by digging coal, as it is called everywhere, seeks a claim certificate on a selected district and when he has obtained it, he himself is tradesman, foreman and collier in one person, works a few hours in the morning as his strength will allow him, takes everything away without measure as long as he finds coal and goes about his other chores in the afternoon. Thus he continues on a daily basis until nothing is left in the upper reaches. And when the costs are felt, both types of tradesman withdraw the hand and let it all become waterlogged (cit. Hue 1910, 373).

Commercial districts in the vicinity of coal pits were able to substitute wood for coal at an early date. The impulse to imitate the British example emanated from there. The salt works of Reichenhall and Lunenburg were dependent on wood for firing for a long time because coal could only be brought to them with some difficulty and much effort. But the situation was different in the salt works at Halle, which could be supplied from nearby lignite fields on the other side of the Saale. Towards the end of the 17th century the physician-in-ordinary to the Elector of Saxony, Nida, promoted the transition to using lignite, which was not distinguished from mineral coal, but met heavy resistance from the salt makers. They feared that coal smoke would make the salt 'useless, unhealthy and wet or even cause bad diseases among masters and servants' (Schulze 1764, 23). Some advantages with coal use were noticed when this resistance had been overcome not only with respect to costs but also the quality of the salt. Thus, it is said in a description of the salt works at Halle of 1708:

> The material for fire making here is either wood or mineral coal. It is noticeable that equally good salt is not boiled with straw, peat or twigs as with wood, but it is even better with mineral coal because it gives a more even fire (Hoffmann 1708, 33).

These examples and travel accounts from England demonstrated that coal could replace wood without disadvantages. In areas where coal was essentially available and no insurmountable transport problems existed, a recommendation to burn mineral coal was a common element in state strategies for wood conservation. In numerous forest orders, apart from the usual calls for saving wood, there are suggestions and regulations for the use of coal. Thus a Saxon 'Item of Resolutions in Wood and Forest Matters' of 28/8/1697 states:

> To remedy the wood scarcity as soon as possible, subjects, in particular black and lock smiths, who can obtain mineral coals, must use them for their needs or at least no longer give them wood or charcoal. (Schmid 1839, 108).

Even in the Saxon 'General Order concerning the Conservation of Woods and Forests' of May 28, 1732 those responsible for the forests are admonished:

> …to find peat and mineral coals, to accustom the inhabitants as much as possible that they use it for their homes and crafts instead of firewood. (Schmid 1839, 139).

Similar regulations were also issued to the *Professiones*, especially blacksmiths, locksmiths, bronze and copper smiths and other trades. It was repeatedly ordered that areas with mineral coal should not be granted wood (see General Order of 2/8/1763 in: Schmid 1839, 163). As early as 1585 the Duke of Brunswick issued an order to use mineral coal for 'smithing, lime burning and brick firing' (Hue 1910, 350).

Here, too, it should be noted that it was considered necessary for the state to order use of coal. If wood scarcity had truly existed, would the consumer not have arrived on his own at using coal instead of wood from considering costs? This was obviously the case in England where no orders to use coal are known. There are two possible explanations: either coal was not yet competitive in all areas or the mentality of economic calculation was not yet developed in Germany. There are indicators for both suggestions.

It has been noted that it lay in the mercantilist and cameralist interest to achieve as favourable a balance as possible in the zero sum game of wealth and power. To obtain a positive trade balance, it appeared sensible to possess as large a domestic workforce as possible. A large population was also desirable for raising taxes and recruiting soldiers. The precondition was that the growing population could be fed with products of the country, because importing food was to be avoided. It was soon noticed that forested land could be used to grow grain if it was possible to establish the use of mineral coal on a large scale.

> Since mineral coal is underground and does not hinder the cultivation of the surface to nourish people, a country that has it requires less woodland and, therefore, may be more populated. This is the main reason why England is much more populated than any other area in temperate climates (Krünitz 14, 1778, 658).

For this reason, promoting coal use was logical in such cases even if no wood shortage existed, i.e. where the expense of a shift to mineral coal burning stood in no reasonable relationship to the price differential with wood. Furthermore, a kind of innovation bonus would have been required for transition to coal. But the slight price advantage was obviously not sufficient to compensate for any inconveniences associated with the transition.

Even in areas where coal could compete with wood, its use only advanced haltingly. A similar resistance to that in England over a century earlier needed to be overcome, but with the remarkable difference that in Germany the absolutist welfare state undertook the task of establishing coal use.

State measures in favour of coal

There are a number of examples of German states actively promoting the employment of coal. Thus, the Bavarian treasury had a quantity of coal produced near Miesbach in 1763/64, which was to be used in lime burning and brick making. For this purpose eighteen brick makers experienced with coal were specially brought from Liège. However, the undertaking failed because Bavarian coal and probably the clay as well had different chemical properties. The bricks did not turn out, and the experiment was cancelled. The coals lay about unused and were offered in vain to black- and locksmiths for use. Finally, in 1765 it was announced that anyone who wished could have them for free. However, consumers did not accept the new fuel (Hue 1910, 344).

With the annexation of Silesia (1742), Prussia had acquired state run coal mines. However, it was difficult to find a market. About two thirds of the area of Upper Silesia was covered in woods, which were often the landlords' main source of income. There was not by any means a shortage of wood in Silesia towards the end of the 18th century, and consumers saw no cause to take upon themselves the inconvenience and uncertainty of firing with mineral coal. The Prussian administration did fear a wood crisis in the near future, which is why it supported coal prospecting in the 1780s and 90s. Frederick II's policy aimed to establish a self-sufficient Prussian energy supply. Because of the difficult transport situation in Prussia, it was cheaper in some areas to import coal from England and Scotland than from the Ruhr district or Upper Silesia. Therefore, imports were forbidden. Around the mid 18th century coal was more expensive in Berlin than firewood, due to high transport costs.

A series of state regulations pursued the goal of establishing coal use. Initially, they targeted the governmental sector itself. Thus, Frederick II of Prussia commanded that garrisons be heated with coal. In 1756 this order was extended to other areas as well. Premiums were paid for a number of uses such as brick and lime kilns, breweries, dye works, distilleries, glass works, bakeries, potash and saltpetre works, not to mention heating.

However, these premiums were often abused. In 1771 the Breslau merchant Christian Ehrenfried Wildner set up a bleaching oven that was fired with coal from Silesian pits. Supposedly, he saved 40% of the heating costs. He received a state premium of 200 Reichsthaler – and then abandoned the oven. Thereupon premiums were again posted and the state offered to pay for the conversion of stoves to coal burning. This, too, was initially unsuccessful (Von der Steinkohlen-Feuerung, 1786). Towards the end of the century Prussia shifted to penalising firewood use in certain areas. 'The Berlin Firewood Comptoir also came to the aid of Silesian coal mining. In 1793 it increased the price of wood and sold mineral coal below the purchase price to promote its use' (Fechner 1901, 496). The Firewood Comptoir was originally instituted as a state organ to ensure the provision of the capital with inexpensive firewood. Now its monopoly permitted the introduction of coal as a primary goal of state economic policy.

The 'Royal Prussian Main Firewood Administration Comptoir', to cite the exact title of the authority, also conducted experiments with different methods of firing. By its commission the fire inspector for building, Heinrich Jachtmann, published in 1794 a 'Treatise on Making Brewery, Distillery and Malting Firings to Save Wood, Mineral Coal and Peat Burning', Jachtmann reported on experiments he made on behalf of the authority in 1780 and 1793 with different methods of firing. Wood-conserving stoves were apparently the most successful. A comparative experiment, in which the municipal brewery of Schweidnitz in Silesia and the Berlin Charité Brewery participated, demonstrated that twice as much beer could be brewed with half as much wood (Faber 1940b, 77). Using coal resulted in no significant reduction of costs compared to firewood, which is why he suggested subsidising the coal price until this fuel had established itself on a larger scale.

Towards the end of the 18th century prices for firewood began to drop, which could be attributed to increased use of fossil fuels. In Berlin the price for a *Haufen* (= 15 m³) of firewood declined from 17 *Thaler*, 12 *Silbergroschen* in 1767 to 15 *Thaler*, 10 *Silbergroschen* (Bernhardt 1874, 68). The artificially low price of coal also pulled down the price of firewood, but coal because of its greater combustion value was much cheaper than wood. In regions in which cheap subsidised coal was offered, firewood increasingly assumed the character of a luxury just as it had earlier in England.

In Silesia the use of coal soon spread in many trades. Thus, a report of 1786 states

> Only a few years ago, its use was restricted to sugar refineries in Berlin and Breslau. Further use in brick making and lime burning, bleaching, dying, brewing beer, distilling and baking and in almost every domestic pursuit was a problem, the possibility of which was long doubted even in parts where examples could be provided. However, currently these doubts, in envisioning both the possibility and the benefit of using mineral coal, have been dispelled most convincingly. Year after year such installations have not only multiplied but due to eager artificing, the urge to imitate and the imparting of better knowledge, especially because of premiums offered by the Royal Silesian War and Domain Chambers, have also increased in distribution, perfection and usefulness (Von der Steinkohlen-Feuerung, 1786, 223).

According to this report, the following numbers of enterprises used mineral coal In Silesia in 1785, partially with state subsidies:

29 breweries
175 distilleries
11 soap works
13 dye works
10 hat makers
5 paper mills
2 saltpetre works
1 potash works

After only two years these figures changed to an astonishing degree. In 1787 the following number of operations used coal in Silesia:

44 breweries
224 distilleries
53 soap works and lime kilns
24 dye works
102 bleachers
16 baking ovens
25 brass and tin smiths
2566 black smiths

67 brick works
64 other trades
2580 domestic stoves

Altogether that made 5,740 burning sites. Such an increase in only two years appears unrealistic. It is clear that either the initial number is too low or the later number too high. Maybe the intention was to create the impression that a rapid acceleration of the substitution process was under way, to motivate imitators. In any case, it is interesting that these figures, which were initially published in the Silesian Provincial Papers (Schlesische Provinzialblätter; 1788, 8, 347f.), were soon copied in the Hanoverian Magazine (Hannoversche Magazin; 25/12/ 1789), where readers were presented with both England and Silesia as illustrative examples.

In other German states, too, governments made efforts to introduce firing with mineral coal. Thus, Prince Wilhelm Heinrich von Nassau-Saarbrücken issued an instruction in 1765 'in what form mineral coals may be properly and beneficially used for warming chambers and quarters' (Haßlacher 1884, 41). It contained detailed instructions on how to adapt the chamber stove for coal fires. Governmental agencies in the Saar District enforced the use of coal from scratch, because it began before 1750 to sell wood from the Saarbrücken Forest to the Dutch on a large scale. Within a short period wood prices had risen so high that the inhabitants had no choice but to use coal. This was in the interest of the state's economic policy since coal mines were run by the state. They saw an opportunity in the increasing scarcity of wood to promote the sale of coal, which was held in little regard. In 1789 an edict was issued by the Palatinate government 'Of the use of mineral coal firing and how to use it beneficially'. Although the Palatinate possessed no coal reserves of its own, it advised the use of imported coal from the Saar District especially in the commercial region around Mannheim, which allegedly could only be supplied with firewood with great difficulty.

Apart from this state support for mineral coal consumption through technical advice from the authorities, the establishment of experimental ovens for bakeries, distilleries, brickworks and other trades, subsidies and premiums of all kinds, preparation of construction plans and the establishment of advisory offices or invitation of foreign experts, there were private associations that undertook similar efforts. Thus, the 'Leipzig Economic Society' sent a master mason to Silesia in 1787, who was to find out what savings might be achieved with coal stoves. The Hamburg 'Society for Promoting the Arts and Beneficial Trades' also called in 1766 for experiments with mineral coal. Finally, a flood of propaganda pamphlets that amounted to as great in volume as the writings on wood-conserving arts must be mentioned. Even the earliest and most widespread publications of this type, the '*Vulcanus Famulans* or Special Fire Use' of Johann Georg Leutmann, which appeared for the first time in 1720 and then experienced a series of editions, juxtaposed both forms of wood conservation as equals:

> There are ... only two ways to counter this deficiency of wood. First, that care is taken to find peat and mineral coals and to introduce their use, in particular that of mineral coals. ... The other way leads to the attempt to manage the use of firewood and to invent machines that heat much with little fire. (Leutmann 1720, Vorrede).

In writings on conserving wood, it was customary to treat coal burning as a special form of wood conservation. We have seen that this was the case in forest regulations. But there was also a special literature about the problems of coal, which increased in volume by leaps and bounds in the 18th century. In a writing of 1802 there is a comprehensive bibliography of publications on mineral coal, lignite and peat, in which a total of 257 titles are cited that are mostly from the German and French speaking region (Voigt 1802). English publications are not mentioned since knowledge of English was not widespread on the continent at that time. It is noticeable that this bibliography only mentions four titles from the period before 1700, while the majority comes from publications of the 18th century. One can justifiably speak of a 'true flood of publications on the topic of alternative fuels' (Gleitsmann 1980, 129), though it should be considered that there was a general increase of published titles in the 18th century and that Voigt was surely more concerned with new than with dated literature.

Resistance and prejudices

Authors who promoted the use of coal in public had to struggle against the widespread perception that it was harmful to health. Many were offended by the peculiar smell. This distrust and discomfort, with few exceptions, was hardly articulated among the literary or even scientific public. It was all the more widespread in the public at the pamphleteering level. This is already evident from the fact that a whole flood of writings was concerned to fight prejudices against coal. Of the few examples of an open expression of such concerns, the article 'Mineral Coal' may be cited, which appeared in 1717 in an economic encyclopaedia:

> They come for most part from England and Scotland, where one cooks with them and heats chambers, but they exude an evil and corrosive smoke that is most dangerous for chest and lungs and without doubt the cause that a certain Englishman proclaims that a third of the inhabitants of London die of the

wasting disease and lung ailments. However, there is a search for the same mineral coals in Germany, too, and especially in Upper and Lower Saxony (Huber 1717, 1555/56).

This passage is found unchanged in the edition of 1762 and if a position like this could still be expressed in a specialised dictionary despite the broad campaign against prejudices, it was probably not perceived to be completely out of bounds. Otherwise, it is writings that are not primarily concerned with the fuel question that are more likely to contain remarks unfavourable to coal. Pamphlets equivalent to those of Evelyn or Nourse cannot be found, but rather formulations that sound remarkably like those of the two Englishmen. Thus, we read about London in the anonymous 'Observations of a Traveler:'

> Since one ... does not burn anything, except for a few sticks used to light the fire, but mineral coals, which appear to multiply dust and darken the air with their black steam as soon as it becomes a little damp... new houses soon become brown and black and look as if they had been smoked. A considerable discomfort that extends to the skin, clothing and especially white laundry and appears to me the chief cause why houses in this country are not plastered white. (*Bemerkungen eines Reisenden* 1775, Bd. 2, 342).

Travel reports like this description of London fed resistance against burning coal. It is not surprising that a century long campaign was required to overcome prejudices. Among the few scientific authors who raised explicit concerns against coal firing was the cameralist Johann Heinrich Gottlob von Justi. We have already met him as a confirmed representative of wood conserving policy and the transition to rational forestry. Entirely in the spirit of mercantilist opinion that the wealth of a country lies in its population, he considers the idea that coal would free land no longer needed to bear forest to be cultivated for grain and feed a larger population. Nevertheless, he arrives at a negative evaluation:

> In the meantime, if it should be true that mineral coals are harmful to human health, as one must conclude from the death registers and their comparison in many towns with respect to the unparalleled higher mortality in London, in Halle and in other towns where almost nothing but mineral coals is burnt; population would suffer on the other hand (Justi 1761, 442).

The foremost task of authors who promoted the new fuel in Germany was proving that coal fumes were not a worry for health. The first writing known to me to praise mineral coal in Germany, the *Sylva subterranea* of Johann Philipp Bünting of 1693, already addressed these concerns and attempted to dispel them by citing classical authorities like Andreas Libavius, Geronymo Cardano, Georg Agricola and Levinus Lemnius. However, he did not deny that Sir Kenelm Digby, a founder of the Royal Society, attributed the lung illnesses prevalent in

England to the effects of mineral coal smoke and that 'English sweat' was considered a most 'burdensome disease' 'caused by the hazardous smoke of mineral coal'. Bünting attempted to relativise the significance of air pollution by attributing consumption in England to a variety of other causes though he admitted that coal burning 'may well contribute somewhat'. Altogether, he named four factors:

i.   Hereditary factors 'because diseases such as consumption etc. *a parentibus transplantantur in liberos,* leave alone that such diseases are *endemii* or *vernaculi* of this country'.

ii. Poor and improper nutrition 'because the English among all occidental peoples have the worst diet by making day night and night day, sleep at noon and booze at night and overload themselves with much meat, fish, spices and wine'.

iii. 'Because frequently poisonous types of mineral coal are found there, whose fumes slightly endanger human health'.

iv. 'Because the English burn mineral coal in open chimneys and sit around them to warm themselves, where many subtle and invisible impurities and effluvia are scattered into the air and communicated to those standing about, which one need not worry about with German stoves' (Bünting 1693, 133f.).

Two years later Friedrich Hoffmann, Professor of Medicine at the University of Halle, produced a treatise that dealt with the effects of mineral coal on health. Hoffmann, one of the most renowned physicians of his age, turned to the academic public, in which the opinion that coal smoke was hazardous to health had been widespread since Digby's warnings. His report, published in 1695, bore the programmatic title *De vapore carbonum fossilium innoxio* (The harmlessness of mineral coal smoke). He reported experiments that he had made with coal and he had come to the conclusion that it contained no dangerous substances. He demonstrated the presence of sulphur but considered it beneficial because it was used in several medicines. The sulphur content of air emitted by coal would rather contribute to purification and was in no way hazardous to health.

> All essentially health promoting vapours can, if they occur in too large quantities become troublesome with respect to health. The like applies to coal smoke if it is too concentrated. In essence, coal smoke is harmless but this does not apply if it is too intensive. In England, where the air is not only rough and salty, but where one eats and drinks alcoholic beverages beyond measure, the breathing air is so much enriched with coal smoke that it damages health by drying the lungs

and the entire body. This must not be attributed to mineral coal as such but to the high concentration of its smoke, i.e. the air is enriched too much with otherwise very healthy substances (Hoffmann 1695, unpag.).

This treatise argues quite carefully and follows the standard argument of the Galenic tradition that the concentration of a substance and not the substance as such is critical. For coal this means that its smoke becomes harmless by dilution. This argument became the most important scientific point of proof in German propaganda writings of the 18th century. Furthermore, Hoffmann could prove in 1716 that burning charcoal in closed rooms could lead to lethal poisoning (with carbon monoxide gas). This frequently led to the conclusion that mineral coal smoke was healthy but charcoal smoke lethal.

In France, too, a number of medical opinions were commissioned to prove the harmlessness of mineral coal smoke (printed in Morand 1777, 33ff.). Therefore, the public could conceive the impression that burning coal was under all circumstances harmless:

> We simply claim that all mineral coal smoke, even in closed rooms, is harmless and that the opinion of doctors, who assume that sulphur vapours inhaled at close quarters are healthy, also applies to resins of the earth among which coal may be counted without dispute (Venel 1780, 67).

The opinion that sulphurous coal smoke cleanses the air and therefore promotes general health was also expressed by Samuel Hahnemann, the founder of homeopathy. In his 'Treatise against the Prejudices against Mineral Coal Firing' published in 1787, he lists a whole series of cases in which the introduction of mineral coal burning supposedly improved the general state of health: in Liège those suffering from 'narrow chest and lung illness' found relief of their complaints; in Valenciennes the introduction of mineral coal firing in 1740 caused chest complaints to become rare; in Lyon, too, they were noticeably reduced. Undoubtedly, there were many with lung ailments in London but that was rather the result of

> Excesses, indoor work, excessive strain of head and body, deficiency, worry and destructive passions ...
> Night life, excess of heating, drink and food, the enormous indulgence of tea, uninterrupted tension and stimulation of the nervous system with all enjoyments possible and the most horrifying and often unnatural passions...(Hahnemann 1787, 14)

were the supposed causes of the bad health of the English – but not coal smoke. Another indicator was that in other regions of England there were as many persons suffering from lung ailments as in London. In Germany, too, he cited examples of the health-promoting effect of coal smoke. Thus, salt-makers in

Halle reached an older age after coal was burnt, epidemics no longer occurred, likewise in Kirn, where the alum works used coal. 'The inhabitants of Dresden, too, where mineral coals are burnt, are healthy and live to an old age' (Hahnemann 1787, 15). Examples of this sort are cited in great numbers to dispel objections against mineral coal.

Hahnemann particularly deplored seeing coal as an inferior substitute for firewood. Instead he advanced its use in wealthy households. They should lead in this matter of highest importance to public welfare and, as enlightened citizens, set an example for the lower orders.

> Also, one should not consider a fuel that is in almost all areas more appropriate than wood as the miserable refuge of the poor since nothing useful is contemptible and industry has always been the most fortunate pride of all nations. The wealthiest houses of Lyon, London, Aachen, Arras, St. Etienne, Liège, Berlin etc. use this fuel by preference (Hahnemann 1787, 5).

There were persons in the educated and enlightened public who were willing to take the inconveniences of coal firing upon themselves since they were convinced that the transition to coal served human progress. As with the introduction of wood-conserving stoves, there were pioneers who experimented tirelessly and spared no efforts or cost to promote further use of the new energy source. They were not interested in personal savings but considered themselves avantgardists of a new technology to which the future would belong.

Such a person was the Hanoverian Alberti, who had become acquainted with coal burning in England and wrote about it in glowing terms. His travelogue reads like a propaganda pamphlet for mineral coal and was probably intended that way. He answered the hygienic and aesthetic objections against the new fuel as if the complaints of Evelyn and Nourse had never existed:

> The strong mineral coal vapour gives London this advantage over all other towns in the world, that one can smell it an hour's distance away. This vapour is at first repulsive to the stranger but he will not sense its unpleasantness any longer if he has spent only 14 days in London. Sir, do not believe that something damaging to the health of a stranger lies in mineral coal vapours of which I acquired ailments as little as others but rather, as a sensible Englishman has judged, it is very beneficial to the inhabitants because it cleanses the air. One should consider that with such a large number of persons and houses built so closely together, of which most have no yards, infectious epidemics could easily arise if mineral coal vapour was not available as an antidote. Thus, Providence cared well for London by forcing its inhabitants to burn coal for lack of wood (cit. Schulze 1764, 34f., likewise also Lindemann 1784, 108f.).

Having returned to Hanover, Alberti built himself a coal stove and attempted to find followers for the new fuel. Over a period of 24 years he experimented with coal stoves and accepted being considered peculiar and a sectarian. Despite all the setbacks to his efforts in effecting further dissemination of the new fuel, he held to his convictions without reservations.

Already in 1766 he had published in the *Hannoverschen Magazin* an 'Attempt to speak on the benefits of using mineral coal fires in domestic stoves'. In it, he picked up the usual reservations rather skilfully and claimed that he used to share them. However, he had 'come to the conclusion that one had to fear a wood scarcity over time' so he had given up his prejudices and entered into the risk of building himself a coal stove. He had only had good experiences with it so now he wanted to publicly encourage the citizenry to follow his example. He would stand with word and deed at the side of any who ventured to undertake the experiment.

This first article obviously did not have the desired effect and there were hardly any imitators. Six years later Alberti reported in the same paper the experiences made after his call:

> Much to my dismay I then found very little following and even a part of the few who followed me tired of this form of heating because the pipes that took the steam and smoke from the stove soon plugged up and their cleaning... brought much dirt into the room. Therefore, they got rid of these stoves at a loss and reintroduced their wood stoves (Alberti 1772, 242f.).

Alberti was not deterred by these setbacks. He undertook a series of experiments and developed a method of cleaning the stoves without making too much of a mess. He now praised this new stove before his audience – not to sell it but solely to make a contribution to the establishment of mineral coal, which he was convinced was necessary for the sake of preserving woods for later generations.

It is apparent from his explications that at that time the post offices, factories, garrisons, police offices, poor houses and servants' rooms were heated with coal stoves. However, he saw it as the duty of the members of the higher orders to lead by example, not just to leave it to the poor to use coal out of necessity.

> This much we know from experience, that the lowly mob is the hardest to win over in all beneficial arrangements but gladly becomes an imitator of great ones (ibid.).

In particular, if the 'great ones' turn to coal burning, the poor would follow. The debate on mineral coal burning was continued over years in the *Hannoverschen Magazin.* Even in 1789 it was considered necessary to publish a question to the public: 'Is heating rooms with mineral coals detrimental to health?' – a question that had found no definitive answer despite years of discussion. Alberti answered the question with the assurance that his state of health had not worsened in the years in which he burnt coal, in fact he had colds less often. Furthermore, he pointed to his experiments to show that burning coal was indeed cheaper than wood.

Looking back in 1790 Alberti reported that the difficulties and resistance related to the introduction of mineral coal were diminishing in Hanover. The improved installation of stoves had made their wider dissemination possible.

> Already 24 years ago there was hope of making mineral coal burning universal, because many tried; but the dirt and fumes associated with this firing scared them all off (Alberti 1790, 153)

– all except for Alberti, who made it his duty to overcome these difficulties.

This example shows that resistance against coal was considerable and that serious efforts were necessary to counter it. The fact that pioneers of the new technology worked in vain or had little to show for such long periods demonstrates that the savings effect could not have been great in a town like Hanover, which was not located near coal mines. Apart from the costs, technological factors played a role that must not be underestimated.

If a stove was refitted for operation with coal, it had to be equipped with a grid to improve ventilation. If one wished to cook or bake with coal, the food had to be separated from the fire. It was not possible to roast a piece of meat over an open coal flame. Nor could meals be cooked in the combustion chamber, as was customary with wood stoves.

Let us remember that the original form of the hearth was an open fireplace in the middle of the hall and that this form of heating hung on in north German farmhouses until the 20th century. In the 18th century it was still widespread. It is obvious that coal could not be burnt on such an open hearth since the smoke would have made staying in the room impossible. In that case the entire house would have to be rearranged to use coal.

A similar problem existed in many middle class and aristocratic houses. Here the flue that was installed over the open hearth and directed smoke into the attic had developed into a mantle chimney. This was a chimney that extended over several floors, with every fireplace in the house having a flue into the chimney. As a rule it was quite large to prevent accumulation of soot, which might catch fire. This chimney was often linked to a so-called 'chimney kitchen', which was often located in the ground floor. This kitchen had no roof but, in a manner of speaking, lay under a turned-over funnel that tapered off towards the top and led into a chimney. The kitchen was only the bottom of the chimney, in which everything was veiled in smoke that could escape upwards but travelled poorly because of the large cross-section.

This chimney kitchen, also known as 'Hell' (Faber 1957, 266), was widespread in large parts of Germany. At times the chamber stoves were heated from the chimney room in this type of construction, so that the rooms themselves remained smoke free. However, being in the chimney kitchen was often very unpleasant, especially during unfavourable weather when the smoke did not escape. It is understandable that in houses of this sort the introduction of coal firing was virtually impossible. Not only did the stoves have to be refitted: the entire house would have to be rebuilt. That coal was resisted under such circumstances is understandable and has nothing to do with irrational prejudice.

Ascendancy of coal

Various efforts to promote the transition to coal burning were finally successful in conjunction with rising firewood prices. Since the close of the 18th century a continuous increase in Prussian coal production can be observed:

Mining Authority District	Year	thousand tons	Year	thousand tons	Year	thousand tons
Upper Silesia	-	-	1790	41	1805	440
Lower Silesia	1740	10	1790	327	1805	875
Province of Saxony	1695	31	1790	55	1805	63
Minden	1770	22	1785	20	1805	31
Ibbenbühren	1747	9	1790	24	1805	56
County Mark (Ruhr District)	1737	116	1790	685	1805	1130

Table 20. Coal production in Prussia

(Hue 1910, 354)

Similar developments occurred in other German states. If and to what extent the use of coal established itself in a particular case depended not least on means of transport. This posed a particularly difficult problem in the Upper Silesian coal district. The Prussian state was unwilling to provide a suitable infrastructure for transport. The task of maintaining roads was left to local landlords and communities, but they had no interest in securing a market for coal with their own means.

Heavy horse-drawn coal transports destroyed recently repaired roads within a short time: 'They were more like a deeply furrowed field than a made road and some wagons had to dump their loads in the bad season to avoid getting stuck' (Schroth 1912, 24). The consequence of these poor road conditions was that continuous provision of more remote locations and consumers was not possible. There some trades suffered from disruptions of production because of a lack of fuel. This led to permanent friction between mines, crafts and landlords, since the latter were not willing to repair roads they rarely used. To alleviate the situation, the 'Coal Mining Support Fund' was founded, which was to finance the construction and repair of roads, but it had no significant success. For a number of metal processing firms it was less expensive until well into the 19th century to use local firewood than to rely on the uncertainties of coal supply and bear the high transport costs incurred in its use. Even in 1846, only 17 of 69 iron works in Upper Silesia used coal or coke, the others continued to work with charcoal (Schroth 1912, 36).

In the Ruhr District, too, coal was initially transported with packhorses and only towards the end of the 18th century were coal roads built, which were usually financed by private companies. They led from the mines into the commercial district of Berg and Mark, to Elberfeld, Barmen and Schwelm. From 1774 to 1780 the Ruhr was expanded for navigation and became the main transport artery. Short tracks were built from the mines to the coal boat docks. In the 1820s longer tracks were built on which horse-drawn coal wagons transported their heavy loads as the precursors of commercial railways.

The inland location of coal pits and commercial centres was the main reason why the replacement of wood by coal began so late in Germany. Only when railways were built could such commercial districts, where transport costs were formerly prohibitive, be supplied with coal. The significant advantages England had because of its insular location and the associated possibility of sea transport is illustrated by the fact that even in the age of railways, in 1886/87, Ruhr coal could not compete in Hamburg with imported coal from England (Crew 1980, 245).

Coal only permanently established itself in the course of the 19th century. Resistance was effective for some time in individual regions, especially since considerable adaptation costs arose in the domestic area. Not only did a fireplace have to be installed, but smoking meat had to be given up. Coal ashes, too, were of little use, while the potash obtained from wood could be processed into soap or sold to glass works. The transition to coal signified another step from self-sufficiency to market dependency for peasant households. It was still necessary in 1839 to take a position against the usual reservations:

> However, if as has been said in certain journals, one calls mineral coal burning hideous because it fouls and spoils everything and even ruins the smell, this can only be in a household where there is little cleanliness... Where there is no pleasant smell, mineral coal can improve it as little as wood (Krünitz 172, 1839, 521).

The fact that existing resistance was rejected in such an offensive tone is a clear sign that opponents were now in a hopeless retreat. Anyone who rejected coal for aesthetic reasons was no longer courted but ridiculed and suspected of a lack of civic virtue, even portrayed as cantankerous and eccentric. In practice, the number of those who rejected coal firing declined. Coal stench was just one of those impositions on behaviour that came in the wake of industrialisation. 'Romantic' resistance against the revolution in daily life associated with industry did not focus on the new source of energy. Protest could pursue other, much more spectacular subjects.

To the lower classes, whose entire life and work was revolutionised by the Industrial Revolution, coal smell was probably a lesser evil – in any case, less of one than a cold home. They were defending themselves against such elementary interventions into their accustomed lives that aesthetic impulses played no role. Members of the middle classes, too, from whom aesthetic and hygienic objections might be heard, found more worthy subjects of complaint. Given the need to overcome the wood crisis, the disadvantages of coal fires appeared a small price to pay. A worsening of environmental quality with the use of this new energy source was finally accepted. Greater resistance was only felt where there were visible infractions against private property, e. g. when smoke damage occurred on vegetation (e.g. Stöckhardt 1850). However, the energy system did not become the symbol of anti-industrial protest like nuclear energy today, even though it was ascribed a symbolic function: smoke stacks became the proud symbol of industrial prosperity.

4. Was the Wood Crisis an Energy Crisis?

It was mainly state initiatives and pressure that promoted the transition from wood to coal in Germany. Among the populace, there was considerable resistance against the new fuel. Does this fact not speak against the customary thesis of wood scarcity in the 18th century? If real wood scarcity had existed, should we not expect that consumers would escape to a more favourable alternative of their own accord? Was not a marketing strategy for coal pits, which at least in Prussia were owned by the state, behind the official attempt to establish coal burning? How else can we explain that it took over a century to establish the use of coal?

One possible explanation might be that the wood crisis of the 18th century was in the first place a timber crisis. The enormous consumption of firewood in combination with agricultural uses of woodlands made it increasingly difficult to find old tree stands that were suitable for construction. If it was possible to reduce significantly the consumption of firewood by conservation, rational forestry and substitution, wooded areas could be used as timber plantations. Since it was much more difficult to replace wood for non-energy uses, the attention of the authorities was directed in the first instance at firewood. Thus, the mining inspector Carl August Scheidt wrote in 1768:

> Wood today is too precious and dear for an innumerable quantity of trunks to be felled in Germany merely for burning, if one can use mineral coal instead, while the wood could be used for necessities where coals cannot (Scheidt 1768, 171).

At first sight there was no shortage of fuel. It was always possible, and with little effort, to produce firewood. During a truly threatening general fuel crisis the supply could have been increased within less than two decades by establishing coppices. Since this was generally known, many contemporaries did not believe in an impending crisis. Firewood was always available at an acceptable price; all the more if one had the hereditary right to draw a wood allowance from the communal forest. In general, it can be said that the fuel aspect was only part of the wood crisis and that part most easily open to a traditional solution.

If, on the other hand, it was possible to solve the fuel problem by using coal, then the woods were unburdened and could be dedicated on a grand scale to raising lumber. This relationship was apparently only insufficiently signalled by the pricing mechanism. It was more profitable for a private forest owner to use woods over the short-term than to establish high forests that could only be sold after generations.

> In some regions of Germany one could see in the last twenty or thirty years how the most unsalutory and reckless mischief was carried out against precious trees and woods and entire forests were felled and entire districts bared because the individual owner can deal arbitrarily with nature. What does he care, who needs money and wants to consume in ten years what shall feed his great-grandchildren, if he leaves a barren and to humans unpleasant, even almost useless patch of land? He wishes to live and they may see how they get by (Arndt 1815, 60).

Every privatisation of public forests and every relaxation of forest supervision led in the early 19th century to a massive worsening of stocks, especially in France after the Revolution. This was also the case in England where not only the fuel supply could not be covered by wood but lumber also had to be imported in large quantities. Here complete liberalisation of land use led to the destruction of forests, which, given the accessibility of coal and the possibility of importing timber from northeastern Europe, had no negative economic consequences.

However, in Germany development headed in the opposite direction. Here direct control of the forest economy through the market could not establish itself since the state authorities suspected that structural deficits would arise if long-term, irreversible processes were involved. Concern for the sustainable existence of natural conditions of production, like social policy, became the sphere of the state, which thought itself more capable of a long-term perspective than private interests. State economic policy adopted a perspective in forestry that went beyond private property. Whoever invested capital in a piece of woodland had to think in time spans of a century or more but that would extend beyond his life span. But the state embraced this position, since it was less concerned with the profitability of alternative uses of invested capital than with the general good.

An important result of the wood crisis of the 18th century was the creation of a scientifically planned and governmentally regulated forestry that primarily sought to produce lumber. The woods were transformed into timber plantations and in the closing 19th century, stocks were far larger than a century earlier. An important precondition for this 'Salvation of the Forest' was the gradual decline of demand for firewood, especially in the commercial sector but increasingly also in urban households. It rested not only upon energy conservation but also on a transition to fossil fuels. We are dealing with an apparently paradoxical result: in Britain the use of coal since the 18th century led to a disappearance of woods except for a small remnant. By contrast, in Germany the woods stabilised and assumed the character that we associate today with the concept of a forest: areas of exclusive timber production with dense stocking.

What appeared to contemporaries as a threatening general shortage of wood was essentially a developmental crisis of the agrarian solar energy regime. In the 18th century two incompatible tendencies collided: the tendency of the solar energy regime towards a stationary state and the technical and economic dynamic that aimed at 'growth'. Similar dynamic phases had existed repeatedly in agrarian civilisations, for example in the Hellenistic Mediterranean, in China under the Sung dynasty or in Europe in the Later Middle Ages. However, such movements and innovations emanating from the social and cultural sphere repeatedly encountered natural barriers. It was not physically possible over the long term to feed these innovation processes: they were regularly strangled by a lack of energy and raw materials.

Europe in the 18th century was in a comparable situation. The agrarian regime once again encountered inherent limits that it could not transcend without a fundamental transformation of its social metabolic basis. Therefore, the much-decried firewood scarcity was only part of a deeper crisis-prone tendency of the agrarian system recognisable in other areas as well. The agrarian mode of production approached a metabolic plateau that it could only surpass by changing its character. European societies of the early modern period exploited the potential of the solar energy regime in a manner without precedence, but there was no compelling reason yet that would cause them to burst its boundaries.

Their limitations become apparent in the fact that nutrition deteriorated steadily from the late Middle Ages, especially in qualitative terms. It has already been noted that under the condition of agrarian production, population growth in a particular area is expressed in people being provisioned at a lower trophic level. When population density grows in a solar energy system, it can be expected that meat consumption will decline in favour of vegetarian food. Precisely this process can be observed in Europe during the early modern period (Abel 1972, 1978; Montanari 1993).

In Germany annual meat consumption in the 14th and 15th centuries was about 100 kg per head. This figure was only reached again in the affluent society of the 20th century. In the 18th and early 19th century, the final phase of the agrarian system, meat consumption was 14 kg per person, i.e. it had dropped to an eighth. This sensational downturn saw the number of butchers in towns decline despite population growth. Of course, it must be remembered that the upper classes continued to consume meat in quantity, so that the general decline in meat consumption meant that the lower classes ate virtually no meat.

Rising consumption of bread between 500 g and 1 kg per day, depending on social status, tended in the same general direction. Almhouses in Paris supplied their clientele with 1.5 kg bread a day in the 17th and 18th century, which may be taken as an indication that bread had become the sole food of the poor. The quality of bread also declined for the mass of the population. The preferred grain, wheat (white bread), was reserved for the upper class, the middle class ate grey bread (rye) and the urban lower class and peasants ate black bread (barley, oats, legumes, acorns, chestnuts). There are many indications of chronic malnutrition attributable to the low protein content of food (cf. Komlos and Baten 1998).

Finally, there were several increasingly severe famines on the threshold to industrialisation. Most were regionally confined but there were years, such as the crisis of 1709–10, in which all of Europe suffered hunger. In France the number of famine years steadily increased in the 18th century. The last of these traditional harvest failure related famines was the so-called pauperism crisis of the early 19th century, which some contemporary observers have attributed to the effects of early industrialisation.

These observations speak for the fact that in the 18th century the preindustrial solar energy system stood at a threshold impeding the growth of important physical parameters (population size, material flow). The energy potential of the given area was in a sense exhausted. Technical advance in agriculture that would improve the efficiency of land use was possible but it always faced the problem of rapidly declining marginal returns. Some examples will illustrate this: the medieval three-field system always left a third of the arable fallow. If the fallow were successfully eliminated with more complex forms of soil cultivation and crop rotation, this would have the same effect as if the cultivated area increased by 50%. However, such an innovation would only be possible once. Furthermore, an increase in land productivity was usually associated with a decline in labour productivity, that is the increased yield of an area was paid with increased labour inputs, e. g. for meliorisation (see Kjaergaard 1994).

This indicates that the solar energy potential of the 18th century was utilised to its limits, at least in terms of then-available agrarian and forest technology. This does not mean that the system was facing an imminent crash, because a solar energy system in principle has the attribute of sustainability, at least in terms of energy. At a defined energy flow level it could have operated permanently. But when land was increasingly in short supply labour pressure would increase and there would be more severe conflicts about the distribution of scarce resources. However, people could always cope with these problems within the framework of the given system through improved methods of land use management.

When the energy potential of the entire surface is utilised, no further growth of the economy or the population is possible insofar as it depends on energy. On the other hand this means that economic growth on the scale associated with industrialisation was only possible because the limits of the agrarian solar energy system were burst in the transition to the fossil energy system. Precisely this constituted the solution that formed the energy basis for the industrial transformation and marked it as a unique epochal discontinuity.

V

Perceptions of Fossil Energy

In 1693 the jurist Johann Philipp Bünting published on behalf of the Electorate of Brandenburg's Upper Court Chamber President a treatise on the potential for using mineral coal instead of wood as fuel. He gave this writing the meaningful title 'Sylva Subterranea Or: Superb Usefulness of the Underground Forest of Mineral Coal/ How the same is granted and imparted by God to the benefit of humans in those places where not much wood grows'. Bünting attempted in this work to ascribe the impending transition to mineral coal to Providence, which as a good Protestant he justified with a reference to Luther:

> Either God's highly enlightened tool Lutherus / or as others will have it / his most faithful supporter and bosom friend Melanchthon is said to have prophesied: That three things will be lacking before the day of Judgement / that is 1. A great lack of true and honest friends. 2. True and solid coin / and 3. Free woods.

This apocalyptic prophecy is marked by that deep ambivalence with which the Christian approaches the end of this world and the dawn of the next, the one in which Christ assumes power. The Day of Judgement closes temporal history with a great cataclysm and must be anticipated with fear. At the same time a new era in the history of salvation begins so that it is also joyful. In any case this is a divinely ordained event that may only be compared to Creation, to Adam's Fall and to the Resurrection of Christ. Revelation assures us that this world will come to an end but no one, including Luther, may know the time, for not even the Son of God knew when he would return.

> As we cannot know the year / much less the day and hour of the final advent of Christ's judgement / we nevertheless know and see visibly / that the same time must be approaching slowly and steadily / because the signs announced by Christ and the Prophets are pointed out and revealed.

Obviously, the world is drifting towards its final stage. This may be seen by many signs, comets and sunspots, turmoil and uproar, finally also in a general failing of nature that may be seen as an expression of its advanced age and approaching death. Nature has become exhausted, tired and barren. Things that were a matter of course have been lost. Humans are smaller and weaker than before,

incidents of monsters and harvest failures are increasing, Christianity is divided
and Antichrist stands at the threshold. That nature itself is dying may be seen
in the fact that something as simple and necessary as wood is becoming scarce.

> For in the beginning almighty nature / or rather the Providence of merciful God
> made for the benefit of the human race a great amount of wild wood / through
> which they did not have to fend against the cold alone / but could prepare meals
> / build / make mines / and use it for other works. But now in these last days of
> the world the great woods have declined so much / that both in warm and cold
> countries little may be found.

The Day of Judgement is near. Lack of wood itself does not cause the end of the
world. Bünting utters complaints: 'how many poor persons must now die
because of lack of wood!' But firewood is not the real problem. Rather, there is
a deeper theological dimension. Lack of wood is only a sign that one historical
era has come to an end and, from the perspective of Christian belief in salvation,
ahead lies an absolute Beyond, a new Heaven and a new Earth. Is the Judgement
near? How should humans react in this situation?

Bünting with his sombre diagnosis of the age is an adherent of apocalyptic thought, which flourished in 17th century Europe (see Harris 1949). The
temper of the age was marked by an immediate anticipation of the Day of
Judgement, and the numerous wars and rebellions indicated that the temporal
history of humanity was coming to an end. In England this view was particularly
held by Godfrey Goodman, the later bishop of Gloucester, who found 'a
barrenness in generall ... the weaknesse of the elements, decay of the heavens, and
a generall imperfection in all things now, in this last old and cold age of the
world' and concluded: 'The generall decay of nature hastens the iudgement'
(Goodman 1616, 367f.).

For Goodman, too, wood scarcity was a sign pointing to the impending
end of the world: 'If wee should commit the like waste in our woods, as formerly
wee have done ... assuredly we should bee left so destitute of fuell, of houses, of
shipping, that within a short time, our land would prove almost inhabitable'
(ibid., 384). However, in the current salvational situation shortly before the end
such waste was reasonable. If the world was going to be destroyed anyway and
there would be no longer an earthly progress of history, a focus on 'sustainability'
or the interests of posterity was completely pointless.

> ... while the world had some time of continuance, when the yeeres were not fully
> expired, then God gave man a minde and disposition to intend the good of
> posteritie: but now in these latter daies, when the world is almost come to an
> upshot, when the period of time is now approaching; no marveile if God leave
> man to himselfe, that out of his own immoderate love of himselfe, neglecting the
> common good, and the good of succession, he should onely intend, in his

buildings, in the waste of his woods, and in all other his actions, his owne private and present commoditie (Goodman 1616, 354).

From this point of view, lack of wood and waste of wood were reasonable because the world was consuming itself as it drifted towards its end. Since preservation of nature was beyond the scope of humans, a measured response to this situation could only be to arm for the final combat and take heed of one's salvation. The crisis is of such magnitude that it is beyond human effort.

Bünting essentially shared this point of view but gave the argument a new twist. He, too, thinks that 'saving nature' is a task that exceeds the capabilities of humans. However, he thought nature can be saved by God with whom humans can side.

> Although a true wood scarcity is present everywhere / so all countries think / how they can fill such scarcity *nolentes volentes* with other fuels. That is not contrary to God / but a way and a means placed into their hands by his omniscient goodness and mercy / so that he may preserve them further until the end of the world.

Therefore, striving to conserve wood and searching for wood substitutes are not contrary to Providence. Humans should not just give in to fate and patiently await the coming of the great cataclysm: they are entitled to search for alternatives and to make themselves at home in this world. And this is where a dramatic turn comes about:

> But now it would appear / as if the Highest again is concerned for us / and will bless us with a *sylva subterranea*, or underground forest of mineral coals.

Bünting claims no less than that the transition to mineral coal is a providential turn. Though he sees the present as *Grundsuppe der Welt*, that is 'this last, worst, most horrible time' (Grimm and Grimm, Vol. 9, 913) that stands shortly before the change into its opposite, salvation, humans can prevent the apocalyptic rupture from happening by co-operating with God. Thus, owing to mineral coal 'further good progress with God's aid cannot be doubted'. The new subterranean fuel moves the desired and feared Day of Judgement further into the future. A new period of worldly progress could begin and the impending total crisis of the present could be overcome.

When Bünting ascribed such providential importance to the transition to coal in 1693, he could only justify this with a traditional theological argument. Contemporaries could not recognise that a route towards a new energy regime was opened. Such a tremendous step away from traditional life styles lay beyond even the most utopian fancy. Europe was on the threshold of the Industrial Revolution and a change of such tremendous extent was about to occur that none could even dream of it. In fact, this process was only possible because coal, a new source of energy, provided independence from the natural limitations of the traditional solar energy system. Humanity possessed a treasure in fossil energy reserves that permitted in a short period of time a tremendous unleashing of forces that accelerated the entire biosphere.

However, when Bünting called coal an 'underground forest' this was meant literally: just as wood permanently grows in above-ground forests that humans may harvest in a sustainable manner, so coals grow in the underground forests. As with all other objects of creation, the following applies here as well:

> That mineral coals, like the other minerals, were bestowed with their special seed by God *in prima creatione* so they will be nourished, multiplied and propagated to the end of the world (Bünting 1693, 46f.).

Bünting follows a basic concept of alchemy according to which there was an analogy between the three realms of creation because they all arose from a single basic substance, *materia prima*. This substance embodied the essence of all three realms (comparable to the divine Trinity) and materialised in the four elements. It is contained in entirety in each element but usually only reveals a particular aspect, i.e. water, fire, air and earth. Every component of the three realms is in a different state of completeness, which is why 'immature' metals can be purified into gold.

All three realms of creation obey the same basic principle. Thus, it is self-evident that metals and minerals including coal grow as stones in their deposits. Therefore, the time that God has granted humanity through access to the underground forests has no natural limit. This is not an foreseeable finite period: the use of coal is as much arranged for permanence as the use of all other natural elements and the end of nature will only come through the direct intervention of God. Therefore, the end of history cannot be calculated through an estimation of reserves and consumption – it cannot be subjected to forecasting as its is in general impossible to predict the date of Christ's second coming. The end of nature is certain but no one can know when it will occur.

1. The Finiteness of Fossil Fuels

However, contrary to all speculation in natural theology, mining experience showed that minerals did not regrow in the bowels of the Earth, but that they are unique, exhaustible occurrences with a finite stock. 'For the shafts finally stop to give metals but the fields always tend to bear their fruits' (Agricola 1556, 3). This also applies to fossil fuels. Therefore, the knowledge that the subterranean forest can only be felled once and that its sustainable use is fundamentally precluded soon established itself.

The fact that the agrarian principle of sustainability could not be applied to coal should have cast a shadow on the enthusiasm with which its use was propagated. People might have recalled an objection levelled against ore mining in antiquity, where a finite stock was used, too: if nature had wished humans to mine metals, she would not have buried them so deeply in the ground. Georg Agricola underscored this argument against mining in 1556:

> The Earth hides nothing and does not withdraw from the eye those things that are useful and necessary to humanity but like a benevolent and kind mother it gives out with greatest generosity and brings herbs, grain, field and tree fruit before the eye and into daylight. By contrast, it cast those things into the depth that must be dug and, therefore, they may not be burrowed out (Agricola 1556, 4).

But a resounding answer to this criticism was delivered: if God had not wanted humans to mine metals, he would not have equipped them with the ability to do so. The fruits of creation were delivered to humans for their enjoyment and that was just as true for minerals underground as for fish in the water. But what if they begin to consume not only these fruits but creation itself? A measure of unease can be felt in contemporary texts that to use up finite resources was to embark on an adventure with an uncertain outcome (see Merchant 1980, 29–41).

Finally, some authors explicitly addressed the problem of how long coal reserves might last. However, an extensive discussion of these questions only arose in the 19th century. Of course, it is no accident that the explicit fear that coal reserves would some day run out was first expressed in Britain. However, it is surprising that the Prussian state, which promoted the shift to coal with a whole series of supporting measures, should have similar concerns. Since it wanted to be sure that this enormous transformational effort was not invested in a short-lived episode, Frederick II requested in 1784 an expert's opinion on how long the city of Berlin could be supplied with mineral coal from Silesia. The Waldenburg Mining Deputation answered that it had enough coal in its district to supply the Brandenburg for over a century with a million bushel (around 50,000 t) annually.

An treatise of 1787 stated that even if coal consumption rose fourfold there would be no lack of fuel for several centuries. In 1800 the minister von Hardenberg inquired of the Superior Mining Authority in Breslau what quantities of coal could be counted upon in Berlin, Potsdam, Brandenburg and Silesia itself. He received the comforting answer that reserves would last another 288 years with a consumption of two million bushels annually (Fechner 1901, 490).

In Britain, too, the fear was already being voiced in the 16th century that coal reserves were coming to an end. As early as 1549 the English Parliament had considered an export ban, though without any result. In Scotland export was in fact forbidden in 1563. In 1610 Sir George Selby testified before the English Parliament that the coal pits of Newcastle would be exhausted in 21 years (Galloway 1882, 53). But even at the end of the 18th century the opinion was widespread that British coal reserves were in fact inexhaustible. In 1789 the Scotch geologist John Williams spoke out against this illusion:

> I have not the smallest doubt that the generality of the inhabitants of Great Britain believe that our coal mines are inexhaustible, and the general conduct of the nation, so far as relates to this subject, seems to imply that the inexhaustibility of our coals is universally held as an established fact (Williams 1789, 159).

He described the colossal waste of energy everywhere and noted that some more accessible coal deposits were already exhausted. The export of coal in particular was subjected to criticism since a country that was poorly equipped with resources was giving up the one advantage it had over rivals.

> The present rage for exporting coals to other nations may aptly be compared to a careless spendthrift, who wastes all in his youth, and then heavily drags on a wretched life to miserable old age, and leaves nothing for his heirs. When our coal mines are exhausted, the prosperity and glory of this flourishing and fortunate island is at an end. Our cities and great towns must then become ruinous heaps for want of fewel, and our mines and manufactories must fail from the same cause, and then consequently our commerce must likewise fail. In short, the commerce, wealth, importance, glory, and happiness of Great Britain will decay and gradually dwindle away to nothing, in proportion as our coal and other mines fail; and the future inhabitants of this island must live, like its first inhabitants, by fishing and hunting (ibid., 172 f.).

If England had to restrict itself to wood as the only fuel source, it would be at a hopeless disadvantage compared to other European nations, 'as most of them have a much greater quantity of wood than we can boast of, or it would be our interest to have' (ibid.). With the later remark Williams is hinting at the fact that England would have to change the structure of its land use at the expense of grain production if there was no more coal. As a countermeasure he suggests forbidding coal exports and to take measures for importing coal from overseas. Curiously, he is mainly thinking of the West Indies and recommends turning them into British possessions before the secessionist American colonies take them over.

The debate triggered by Williams's warnings in England found a continental echo. In Hanover, which was governed by the English king in personal union and looked more towards England than other German states, this theme was picked up in the same year. Thus, an article appeared in the *Hannoversche Magazin* (Hanoverian Magazine) in 1789:

> Many fear that all coal pits in England will finally be completely exhausted and that country will suffer complete lack of coal. The quantity of coals dug up every year is indeed considerable; and because coals do not regrow the time must come when they are entirely used up and no more mineral coals are to be had anywhere. This dismal time is still far in the future and the lack of mineral coal will not be a concern soon (Etwas über den Gebrauch und Verbrauch der Steinkohlen in England, 1789, 122).

After this debate had calmed down, Hanover could conduct the replacement of wood by coal unperturbed. Subsequent political events and the associated separation of England from the European market meant that the problem of coal resources and possible export restrictions played no significant role until the 1820s. Only in geological literature were further calculations carried out with respect to the size of the deposits, but these questions were no longer matters of public discussion.

This problem only gained new currency in the 1820s when the House of Lords was called into a conflict between coal mine owners and London coal merchants and a Parliamentary Committee was called to prepare a legal settlement. The background to the conflict was the fact that coal was not taxed by weight but by volume in London. This made collecting taxes easier: only the baskets in which coal was carried ashore needed to be counted.

The merchants, who bought a quantity of coal on board of the ships and, therefore, paid taxes, harboured the suspicion that it was more favourable to take only large coal pieces because then the weight was more favourable in relationship to volume. However justified this assumption may have been, it had as a consequence that smaller coal pieces could not be sold, so there was no point in taking them to London. Coal was already graded at the mine, leaving a third unsellable on the waste pile to be used for the operation of drainage pumps, or to simply be burnt as waste.

The mine owners now wished coal to be taxed solely by weight because then the sellable part of production could be increased. To achieve this they developed the following line of argument: the prevailing taxation practice promoted wasting coal and since reserves were finite, they were used up more rapidly. Thus, it was in the national interest of England to change taxation methods. The coal merchants countered that deposits were so huge that no conservation was necessary.

Both parties now needed to support their position with expert's opinions. Coal merchants commissioned opinions that resources were as good as inexhaustible; the experts of the mine owners were required to show the opposite. The first round before the House of Lords went to the coal merchants. The expert Hugh Taylor testified that British coalfields would still last over 1,727 years (House of Lords 1829, 77).

The committee of the House of Commons undertook a more thorough inquiry and appointed the geologist Adam Sedgwick as an independent expert. He refuted Taylor's calculations and based on current consumption, i. e without assuming future consumption increases, arrived at 300 to 400 years until British coal would be depleted. In particular, he rejected the assumption that increased labour inputs, improved prospecting and access to deeper layers would extend yield ad infinitum, as the commission members arguing in analogy to agricultural production suggested. Sedgwick emphasised the principle difference in dealing with finite and renewable resources:

> If encouragement would cause the production of beds of coal as of annual crops of corn, it would be so; but as nature has limited the quantity of coal, and any reproduction of it is impossible, if you increase the consumption the total exhaustion will be accelerated (House of Commons, 1830, 246).

The committee members were accustomed to thinking in terms of agrarian production and the conventional solar energy system within which permanent and sustainable yield increases were possible. By contrast, Sedgwick emphasised the fundamentally different character of consuming a fixed stock: here technical progress will only lead to an acceleration of consumption:

> I only wish to press upon the Committee my perfect conviction of the exhaustability of the coal-field, and the fallacious opinion which prevails respecting this coal field and all coal fields, that by the simple operation of digging deeper in those districts marked as carboniferous, you will certainly find other beds of equally good coal; such is not the fact (House of Commons 1830, 246).

A series of expert's opinions mostly prepared by geologists were cited before the committee and, apart from a few exceptions, did not differ much. The mine owners' cause was finally successful and the taxation method was changed by law. However, after the problem had made it to the public in this specific context, a debate was launched that would last for most of the century. A multitude of prognoses was established of which the most important are cited in Table 21.

To the public, which thought in categories of political economy, the finite nature of coal deposits was a serious challenge. The realisation of the problem was complicated further because the assumption that coal reserves were small suggested prevention of coal exports was in the British national interest, just as Williams had already demanded in 1789. At this time, when the Napoleonic wars had ended, public debate was raging on the abolition of the

Author	Year of the Prediction	Years until Exhaustion
H.G. MacNab	1792	360
J. Bailey	1801	200
Thomson	1814	1000
R. Bakewell	1828	200
		(+ Welsh deposits: 2000)
H. Taylor	1829	1727
A. Sedgwick	1830	300–400
W. Buckland	1830/36	400
Greenwell	1846	331
T.Y. Hall	1854	365 or 256
W. Fordyce	1860	300
E. Hull	1864	450
W.S. Jevons	1865	110
P. Williams	1871	324–1695

Table 21. Predictions of anticipated life of British coal reserves
(Sources: authors cited; House of Lords 1829 [Taylor]; House of Commons 1830
[Sedgwick]; Holland 1841 [Bailey, Thomson]; Fordyce 1860 [Greenwell, Hall];
Jevons 1865 [Hull]; House of Commons 1871 [P. Williams])

Corn Laws, with arguments grounded in economic theory. These laws were
introduced during war for reasons of national independence with the consequence that grain prices and consequently land prices and rents rose. After the
end of the war demands were made to abolish these laws so that cost of living
and nominal wages could be lowered. Landowners, for obvious reasons,
opposed this deregulation of grain imports but, since grain tariffs were a
Parliamentary matter, they had to defend their interests with general theoretical
arguments.

In this context two camps affiliated with two economic positions formed
in political economy. On the one side stood the landowners' spokesmen, who
used Malthusian and underconsumption arguments and supported grain tariffs;
on the other side were industry's spokesmen, who promoted absolute free trade
which would lower grain prices and hence wages of labour. This confrontation
became one of ideological principle. Representatives of classical political economy,
especially David Ricardo, John Ramsey McCulloch and Richard Cobden, saw

free trade as an economic cure-all that would bring wealth, growth and freedom. To avoid undermining the free trade principle, classical economists tended to play down the problem of finite coal reserves and gave credence to favourable estimates. They did not want to sacrifice the free trade principle for the sake of this marginal question and hand an argument for continued import restriction on grain to landowners.

> It is the possession of her coal mines which has rendered Britain, in relation to the whole world, what a city is to the rural district which surrounds it, – the producer and dispenser of the rich products of art and industry. Calling her coal mines the coal cellars of the great city, there is in them a known supply which, at the present rate of expenditure, will last for 2000 years at least; and therefore a provision which, as coming improvements in the arts of life will naturally effect oeconomy of fuel, or substitution of other means to effect similar purposes, may be regarded as inexhaustible (McCulloch 1832, 268).

Not only does McCulloch extend Taylor's estimate by 2,000 years, he also simply declares the deposits to be inexhaustible if substitution and savings are considered. This enables him to maintain his strict free trade position with respect to finite raw materials:

> It is, therefore, quite idle either to prohibit, or impose heavy duties on, the exportation of coal, on the ground of its accelerating the exhaustion of the mines (ibid., 271).

McCulloch could hang on to the free trade principle and permit the export of coal but only with a more or less conscious self-delusion. Other contemporaries, who were as a matter of principle in the free trade camp but did not permit themselves this wishful thinking, took a restrictive position with respect to coal export. Among them was the famous geologist William Buckland.

> We are all fully aware of the impolicy of needless legislative interferences; but a broad line has been drawn by nature between commodities annually or periodically reproduced by the Soil on its surface, and that subterranean treasure, and sustaining foundation of Industry, which is laid by Nature in strata of mineral Coal, whose amount is limited, and which, when once exhausted, is gone for ever. As the law most justly interferes to prevent the wanton destruction of life and property; it should seem also to be its duty to prevent all needless waste of mineral fuel; since the exhaustion of this fuel would irrecoverably paralyze the industry of millions (Buckland 1836, 537).

Buckland differentiated systematically between renewable and non-renewable natural goods and wanted the state to treat them differently. His argument followed a quasi-natural-law line. Just as John Locke disputed the legitimacy of private property in cases in which goods are destroyed which are required for subsistence, Buckland demanded state intervention when waste of exhaustible resources must be prevented.

> The Tenant of the soil may neglect, or cultivate his lands, and dispose of his produce, as caprice or interest may dictate; the surface of his fields is not consumed, but remains susceptible of tillage by his successor; had he the physical power to annihilate the land, and thereby inflict an irremediable injury upon posterity, the legislature would justly interfere to prevent such destruction of the future resources of the nation (ibid.).

As a countermeasure he suggested restriction of exports and energy conservation. However, energy conservation is not a separate source of energy but can only effect that available reserves are consumed more slowly so that the time until final depletion is extended. The logic of consuming a fixed stock is fundamentally different from the logic of handling renewable resources. The realisation of this problem was a serious challenge to economic thought. No less than a departure from the paradigmatic basis of the solar energy system was required. And in the end the problem of finite energy resources emerged as central to the very historical destiny of the industrial system.

2. Classical Political Economy and the Stationary State

Classical political economy, which had been elaborated from Adam Smith to John Stuart Mill, rested as a theory entirely on the paradigmatic foundations of the agrarian solar energy system. Its epochal significance lay in the fact that it gave the principles of this system a theoretical definition just when it was in the process of transforming into the new fossil energy regime. The theoretical construct of Smith, Malthus, Ricardo and Mill is therefore characterised by a peculiar anachronism that was even shared by its harshest contemporary critic, Karl Marx: it ignored the material and metabolic foundation of industrialisation and in the end remained within the conceptual frame of agrarian society, which it nevertheless analytically penetrated in an unprecedented manner (see Wrigley 1994). Classical political economy was a theory of solar energy flows. It revealed the foundations of the system at a time when it was in a state of dissolution. Therefore, its prognostic power, for example with respect to the development of population and material standards of living, was low and later observers found it easy to refute its predictions. However, if the fossil energy regime is merely an episode in world history, the assumptions of classical economy may regain a new relevance.

Classical economists essentially differentiate three factors of production: labour, capital and land. Labour is a transforming activity that modifies objects and in that sense it is productive. It is fundamentally bound to humans and its supply in the market is closely related to the population level. Since humans supposedly have a basic ability and a natural tendency to propagate, labour is a factor that can never become scarce. To the extent that sufficient food is available, humans will multiply and they will repeatedly reach the limit of food availability. Wages as the price of labour are in the end determined by the cost of food and the means of subsistence, and they tend to find an equilibrium around the minimum for subsistence at which workers can just maintain themselves and their offspring.

Capital is created by abstinence from consumption, i.e. by saving, and, therefore, is an accumulation of past labour. In principle, there should be no limit to the accumulation of capital because every amount saved could be augmented by an additional amount. However, if the yield of savings, i. e. the profit of capital tends towards zero, capital formation stops. Savings are not worthwhile any more because every consumption abstained carries the risk that future consumption may not be possible.

The third factor is generally called land but essentially it means the 'natural agent' in production (Mill 1848, 1,X, §1). Therefore, land is the collective term for reproducible scarce natural factors. In contrast to labour and capital land cannot be increased without limit because its extent is predetermined by nature. In the end, land forms the stable framework of the factors of production and its availability sets a final limit: 'This limited quantity of land, and limited productiveness of it, are the real limits to the increase of production' (Mill 1848, 1, XII, §1).

Thus, the natural factor land forms a limited stock while labour can increase through population growth and capital through accumulation. Since not only food but also the raw materials of commercial production come from land, it must be anticipated that ever greater amounts of labour and capital must be invested in working the soil. The attempt must be made to obtain higher yields through increased effort. This striving is opposed by an important observation that forms the iron foundation of classical economics:

> Each equal additional quantity of work bestowed on agriculture yields an actually diminished return, and, of course, if each additional quantity of work yields an actually diminished return, the whole of the work bestowed on agriculture in the progress of improvement, yields an actually diminished proportionate return (West 1815, 6f.).

The consequence of this observation is that there can be no unlimited expansion of production and therefore no limitless growth of population or capital accumulation, 'for the land being limited in quantity, and differing in quality, with every increased portion of capital employed on it there will be a decreased rate of production, whilst the power of population continues always the same' (Ricardo 1821, cap. 4). Cultivation of the soil reaches an extensive limit at the point where the cultivation of one additional soil unit yields less than the additional input of labour or capital costs. An intensive boundary also exists. An increase of capital and labour inputs initially allows the marginal yield to grow to a maximum. When it is reached the marginal yield continuously drops until a further increase of inputs results in no further addition to the yield. At that point the stationary state of agriculture and material production in general is reached.

That essentially fixes the course of the economic process. With the help of capital labour can produce goods on land and with products of land, but the economic process cannot emancipate itself from the confines of nature. There is no long-term tendency of 'economic growth' but the level of production oscillates around an average: 'the same retrograde and progressive movements, with respect to happiness, are repeated' (Malthus 1826, I, 16). Thus, there can be phases of expansion (progressive states) and phases of decline (declining states), but the overall process finally enters a stationary state (Smith 1776, 99; compare Malthus 1826, II, 58). These oscillations with creeping elevation of levels are the logical expression of a solar energy system in which it can always be attempted to approach an upper limit asymptotically.

In David Ricardo's theory of rent this conceptual pattern becomes particularly clear. According to Ricardo the value of a good consists of the components wages of labour, profit and rent with the value of the good being measured by the work expended. The amount of wages is inversely proportional to profits, because the value of the good results from the aggregate costs of production. Whatever is received as price above the wages paid can (after deduction of fixed costs) be appropriated as profit, with the amount of wages in the last instance being determined by the labourer's costs of living, which fluctuate around the subsistence level. Since the labourer's most important food is grain the nominal wages and in the end profit, which is coupled to wages, depend on the level of the price of grain.

The price of grain is determined by the costs of grain production, including rent, which in turn is a function of the fertility of various soils. The most infertile soil that is still cultivated bears no rent. Since a particular quantity of grain will fetch the same price without regard for the quality of soil on which it is produced, an additional profit arises from grain that was grown on better soils and will in the end flow to the landowner. Every soil with higher fertility than the worst soil that is cultivated yields a rent to its owner that is equal to the respective differences in costs of production. Natural quality differences of soils are reflected in different rent levels in this manner, since the price for the same quality of grain must always be the same independently of how fertile the soil is from which it came.

A historical process can be deduced from this basic consideration. If population grows, either poorer soils need to be cultivated or ever greater quantities of work and capital must be employed on cultivated soils to increase yields. In the course of this movement the quality differential from the best to the worst soil progressively increases, and that means the rent differential also grows. According to Ricardo's assumptions the nominal wage will rise with the rent increases and the grain price, wherefore by definition the profit rate of capital declines. If it tends towards zero the stationary state is achieved in which no further accumulation of capital and, therefore, no economic growth will take place. Since the demand for labour cannot increase further and wages lie at the subsistence level, population growth must stop. In the stationary state all but the landlords live in misery, the world has reached its maximum population and progress has come to an end. This tendency towards a stationary state is essentially unavoidable even though it cannot be foreseen when this state will be achieved.

The law of diminishing marginal returns also fundamentally applies to manufacture to the extent that it relies on processing agricultural products.

> The materials of manufacture being all drawn from the land, ... the general law of production from the land, the law of diminishing returns, must in the last resort be applicable to manufacturing as well as to agricultural history (Mill 1848, 1, XII, §3).

There are a number of tendencies that counteract the development towards a stationary state. The most important is technological progress, i. e. an improvement of productivity. In agriculture this results in a piece of land yielding more without incurring additional (labour) costs or the yield remaining the same while costs decline. In commercial production it decreases the price of goods that are either a component of the cost of living or are employed in agricultural production. In particular in the commercial sector progress may be expected because natural conditions are more open to human intervention and the continually declining marginal return can be compensated by a 'continually diminishing proportional cost'. If the tendencies are combined, it may be anticipated that more labour and capital will be directed towards agriculture while productivity in factories rises, so demand for labour will rather decline there. Furthermore, it can be anticipated that prices for food will rise while those for clothing, shelter etc. decline so that the overall cost of living will remain stable.

Free trade has a comparable effect to technical progress since it leads to the import of grain that comes from areas in which marginal lands are not yet cultivated. That market forces should be allowed a free reign follows in particular from the demanding tendency towards a stationary state dictated by natural conditions. Free trade and the free market economy are the best methods to extend the time until the stationary state is reached.

England in the early 19th century is in a 'progressive state' – that far the economists are in agreement. It is also certain that this state cannot be permanent: 'The increase of wealth is not boundless.' At some point the stationary state will be reached and 'all progress in wealth is but a postponement of this' (Mill 1848, 4, VI, §1). But the natural limits set to economic growth are flexible and can be expanded:

> The limitation to production from the properties of the soil, is not like the obstacle opposed by a wall, which stands immovable in one particular spot, and offers no hindrance to motion short of stopping it entirely. We may rather compare it to a highly elastic and extensible band, which is hardly ever so violently stretched that it could not possibly be stretched any more, yet the pressure of which is felt long before the final limit is reached, and felt more severely the nearer that limit is approached (Mill 1848, 1, XII, §2).

Therefore, the limits of growth set in the solar energy system prove to be ever expanding limits. The system is highly elastic and in principle it is possible at every level to push the boundaries a little further. We can illustrate the logic of this argument with a technical example. Let us assume we have a stream with a certain drop and a certain quantity of water that flows past a particular location. The quantity of energy theoretically available at this site is constant. The energy flow will be tapped with a water mill. We can imagine that in the course of a historical process several mills with increasing efficiency are built to tap into the energy flow and redirect it for human purposes. Different technical principles, methods of construction and degrees of intervention in the energy flow will be employed. In a conceivable final state the entire streambed has been laid out in pipes and set with highly efficient turbines, whose degree of efficiency will be close to what is physically possible.

Each of the mills that are built pushes the limits of energy use up a little. It is not very probable that great leaps will be made during this historical process. Rather, a series of individual growth processes that each achieves a temporary stationary state are involved. We do not have a continuous 'exponential' growth curve before us, but rather the millwrights feel their way towards a sequence of asymptotes within the given state of technology. Instead of steady growth we have growth cascades, i.e. phases of more rapid growth when a new technology is applied and almost unnoticeable fine tuning processes when a familiar technology is exhausted. In the end there will be a final asymptote that is determined by the natural limits of mechanics and can no longer be budged by technology. When this limit has been approached by 95%, no more great breakthroughs are possible. Therefore, there is no continuous and infinite expansion of growth limits but only an infinitesimal approach to a definite limit. This is precisely the economic logic of the solar energy system.

Another characteristic is important. The stationary state of classical economists is marked by a purely reproductive management of material flows at a very high level. A yield limit for soils is reached but production can be continued permanently at this level. When the limit is reached there is no turning point. The mill presented above can work as long as water flows and any mill on any technical level can do that! The amount of energy available altogether over long periods of time is infinitely large; what is limited is only the amount available over a particular period of time. The agrarian solar energy system is, as far as its energy base is concerned, set up permanently and each asymptotic level that is attained essentially can be maintained forever. These are characteristics that fundamentally differentiate the solar energy system from the fossil energy system.

If population size can be successfully stabilised below the final maximum through contraception such a stationary state can be devoid of terror and contain rather pleasant traits. Such a submaximal stationary state of population and agrarian production, as John Stuart Mill conceived it, is at the same time compatible with an expansion of industrial production if capital can be provided. If an increase in real wages does not immediately lead to an increase in population and thus consume itself, as Malthus and Ricardo suspected it would, a long-term improvement of the standard of living can occur with increases in productivity. In this case, too, free operation of market forces is reasonable.

The stationary state, whether it is the sombre version of Malthus and Ricardo or the optimistic view of Mill, is an expression of the paradigmatic structure of the solar energy system. In such a system the goal is always to tap more or less skilfully into an energy flow. Increases of productivity must be understood as increases of energy efficiency. In Malthus's version they permit feeding more people on a given area or, in Mill's version, to feed and provide the same number of people better. Within this process technical progress has the function of raising the level that can be permanently managed. Therefore, the limits of the stationary state are flexible, as Mill pointed out. It is possible to increase energy efficiency at any level through skilful intervention or to optimise the distribution of energy flows that are tapped. The economic mechanisms in the formation of prices, wages, profits and rents can optimise themselves in a free interplay of forces. Even if this leads to temporary imbalances, they are never irreversible processes: the soil that yields the harvest is as indestructible as the sun from which the energy supplying the system comes.

3. Jevons and the Contraction of the Industrial System

With the finiteness and exhaustibility of fossil fuels, a factor with completely novel features entered the economic game. The logic of coal is fundamentally different from the logic of solar energy. With its use the economy separated from the agrarian base that had been the foundation of physiocratic and classical economic theories. It would take some time until the implications of this process were recognised in all their consequences (see Martinez-Alier 1987). The limit set by the exhaustibility of fossil deposits has a completely different character than the growth limit of the solar energy system: no stationary state is possible based on fossil energy; when this system has reached its limits, a new contraction must set in. The only escape could be a shift to novel energy carriers. Therefore, the fossil energy regime is a transitional regime and the society built upon it is a transitional society.

It was William Stanley Jevons, a founder of marginal utility theory with his *Theory of Political Economy* of 1871, who in *The Coal Question* of 1865 was the first to recognise the problem in its entire complexity and significance. First, he turned to the prognoses. The time period until the exhaustion of coal deposits depended on two independent factors: the extent of reserves and anticipated consumption. The difficulty of prognosis lay in the fact that the coal stock might be known but not future consumption. Jevons' predecessors had used two different procedures in estimation. Either they extended current usage rates into the future, which extended the durability of reserves, or they extended the absolute increase rates of the past decade in a linear manner.

Jevons, however, assumed an exponential growth of coal consumption. First, he demonstrated that the use of all significant physical variables had increased 'in a geometric series' during the past decades. That applied to population numbers, coal consumption, iron production, wood and cotton imports, the export of industrial goods and shipping tonnage. Thus, he arrived at the following conclusion:

> ...that our trade and manufactures are being developed without apparent bounds in a geometric, not an arithmetic series – by multiplication, not by mere addition – and by multiplication, too, always in a high, and often in a continuously rising ratio (Jevons 1865, 191).

From this observation and the current trend of coal consumption, he concluded that the annual growth of coal production lay about 3.5%. Now calculation with the exponential function had revealed in the course of the debate on Malthus's population law that 'geometric' growth leads to such huge numbers after some years that any possible resource base smaller than infinite must soon be exhausted. Therefore, it did not matter how large British coal reserves were – the time when they would be exhausted could not be too far off. The question of deposit size was of only secondary importance. Jevons concluded:

> That we cannot long maintain our present rate of increase of consumption ... that the check to our progress must become perceptible considerably within a century from the present time ... that our present happy progressive condition is a thing of limited duration (Jevons 1865, 215).

He estimated on the basis of his assumptions that coal reserves would be exhausted in about 110 years and assumed that there would be no effective remedies. Jevons' prognosis of an exponential increase of British coal consumption agreed with reality until 1910 (Varchmin and Radkau 1979, 114). Only the advances of mineral oil and hydroelectric energy caused coal consumption to decline. The peak of 1910 was never again achieved.

The publication of this prognosis caused a stir in Britain. Among other business, a commission of the House of Commons appointed in 1871 looked at this estimate. They had to admit that the predicted figures had been met over the past six years but doubted that that would continue to be the case. In particular, they noted the parallel to the Malthusian population prediction, which was accurate in the first half of the century, but no longer in later years. The expert Price Williams assumed that the rate of increase of coal consumption per capita would decline just as the rate of population growth had. Beginning in 1941 consumption would be stationary. Under these conditions he arrived at 2231 as the year of exhaustion, 360 years after the present. The British population would then lie around c. 140 million. Under consideration of unconfirmed resources he estimated the following intervals until the exhaustion of British coalfields:

- 324 years at exponential growth;

- 433 years at additive increase;

- 1695 years at constant consumption.

But even the Parliamentary Commission had to admit that Britain's economic superiority could not be maintained since, in the end, it relied upon a monopoly of cheap energy (House of Commons 1871, XVIII). The debate was not concluded at the time. A revival of interest came with the declining productivity of British coal mining towards the end of the 19th century due to the exhaustion of more accessible deposits, while it continued to rise in the Ruhr district. Bad as this may have been for the energy basis of British industry, a stagnation of consumption also resulted from it. Thus, a Royal Commission appointed in 1901 was able to criticise the assumption of constant rates of growth and suggested handling coal more sparingly in the future. An increase in the price of coal could bring the country closer to this goal (see Jevons 1906, preface by A.W. Flux).

Jevons considered energy savings through improved energy efficiency possible in coal use, but did not believe that the overall coal consumption could be decreased that way. He estimated that private households, iron production and other branches of industry used British coal around 1865 in equal shares. Since he assumed that the demand for energy by private households (heating, cooking, lighting) increased roughly at a constant rate, scope for conservation through improved efficiency existed. In factories this was exactly the opposite. If energy was conserved there, production costs would decline, leading to a price reduction and, in turn, increased demand. Overall production rose so that a reduction of energy costs per production unit would cause more to be produced and on balance more energy would be used. He proved this with the case of the steam engines: the better the degree of efficiency, the more steam engines were built, in other words, the more coal consumption would rise. By conserving energy through an increase in efficiency, reserves could not be extended; on the contrary, conserving energy in this manner might shorten their life.

In popular science writings of the 19th century there was a widespread faith in the unlimited possibilities of technical progress that would eventually transgress the limits of provisioning with fossil energy. Jevons discussed each offered or projected energy source such as wind, natural electricity (lightning, Northern lights), water power, the tides, wood and finally petrol and concluded that none could replace coal – petrol, for example, because supplies were even more restricted than those of coal. If coal were to be replaced by wood a land area twice the size of Britain would be needed, according to his calculations.

> Among the residual possibilities of unforeseen events, it is just possible that some day the sunbeams may be collected, or that some source of force now unknown may be detected. But such a discovery would simply destroy our peculiar industrial supremacy. ... We must not dwell in such a fool's paradise as to imagine we can do without our coal what we do with it (Jevons 1865, 145).

England's special role as the workshop of the world rested solely on easily accessible coal. Once this advantage was gone, England would lose its superior position in the world market. It could not import coal since it was the basis of the British trade balance. The producing countries would have such locational advantages due to transportation costs that they would rather establish their own industries than export coal. Above all he was considering Germany and the USA, the feared main rivals in the world market.

So far the emigration of the British population surplus into the colonies had caused exportable grain to be produced there and the creation of a market for English manufactured products. Thus, England was provided with raw materials (other than coal) and exported industrial goods in exchange. That would change the moment the (former) colonies had established their own industries. To at least stave this off, coal must not be exported or its export should be burdened with such high tariffs that the comparative advantage remained.

Overall, Jevons counted firmly on England's industrial growth coming to a standstill and reaching a turning point. The subsequent industrial decline of England was unavoidable since no stationary state was possible based on fossil energy. The growth barrier based on limited reserves of raw material, especially the key resource energy, is fundamentally different from the barrier that Ricardo and Mill saw in the tendency towards a stationary state.

> A farm, however far pushed, will under proper cultivation continue to yield for ever a constant crop. But in a mine there is no reproduction, and the produce once pushed to the utmost will soon begin to fail and sink to zero. So far, then, as our wealth and progress depend upon the superior command of coal, we must not only stop – we must go back (Jevons 1865, 155).

The social and economic order that temporarily rested upon a superabundance of fossil energy would not be transformed into a stationary state. A sustainable economy based on coal was impossible. In foreseeable time a contraction would be initiated. The free flow of market forces would not develop counter tendencies to coal consumption. On the contrary: free trade would cause its export and 'conservation' – efforts to use energy more rationally by increasing efficiency would finally result in consumption growth, since energy supplies produced that way become more affordable. In contrast to the model of Ricardo and Mill, development would not lead slowly and relatively painlessly to a stationary state at a level that could be reproduced sustainably and permanently.

In the framework of the solar energy system, with the developmental logic described by Ricardo and Mill, the *marginal return* will decline to zero in the course of time, i.e. the supply curve will approach the asymptotic upper limit of a sustainable resource flow. In the framework of the fossil energy system, whose logic was first analysed by Jevons, the *entire return* will decline to zero at some point in time, i.e. the resource flow as such will cease. When this limit has been reached, other physical variables must retreat as well, especially population numbers and the material standard of living. The central difference between solar and fossil systems lies in the fact that visible negative feedback processes exist in the former because every growth process causes changes that hinder future growth as marginal returns decline. By contrast, in the fossil energy system positive feedback exists so that long-term growth is possible, albeit if not permanently.

From this perspective, industrialisation and economic growth are only transitory. At that time Jevons foresaw the looming adaptation as so near in the future that he recommended paying off England's national debt as long as means were still available. Under current conditions of economic prosperity this would be less difficult than when economy and population were beginning to shrink.

Retreat to a more modest state was unavoidable. Jevons comforted himself with the thought that England would have achieved a unique position in world history when this wealth was gone. Therefore, the English should advance on the path of industrialisation, since this alone could enhance the British contribution to the advance of humanity. But demise from the stage of world history was unavoidable. So achievement and triumph should be grasped for as long as possible:

> Britain may contract to her former littleness, and her people be again distinguished for homely and hardy virtues, for a clear intellect and a regard for law, rather than for brilliancy and power. But our name and race, our language, history, and literature, our love of freedom and our instincts of self-government, will live in a world-wide sphere. ... If we lavishly and boldly push forward in the creation and distribution of our riches, it is hard to over-estimate the pitch of beneficial influence to which we may attain in the present. But the maintenance of such a position is physically impossible. We have to make the momentous choice between brief greatness and longer continued mediocrity (ibid., 349).

So there was an alternative to Jevons' favoured historical perspective of short-term national greatness, the continuation of growth to the point where scarcity of raw materials brings it to an abrupt halt. This alternative of 'longer continued mediocrity' consisted in avoiding everything that tended towards exceeding stationary possibilities. This would have the advantage of limiting the height from which the system would of necessity fall before reaching a stationary state. But Jevons could not muster enthusiasm for such a prospect. England's place of honour in the book of history was more important to him than a policy of a 'soft landing' that intended to voluntarily fix life at the level it would be reduced to sooner or later.

The finiteness of the fossil energy system set a natural barrier to the capitalist growth economy that could not be as readily mastered by the mechanisms of the market economy as the classical natural barrier resulting from a combination of population growth and declining marginal returns in food production. The latter could be mitigated by letting the free interplay of forces run its course because this would provoke the optimal reaction to this barrier: the progress of productivity. The principle of self regulation does not function optimally with resources that are not only finite but of which consumption is irreversible. Since economic subjects only think in limited time frames, significant savings only become important when scarcity is acutely felt. A predicted scarcity in the far distance is not sufficiently signalled by price mechanisms when current supplies are large.

Therefore, a level of human populations and material flows of goods can arise that cannot be maintained without fossil energy. When the market finally enforces adaptation, it may be far too late. The irritation of a market economy theoretician at this fact is clearly felt in Jevons' writing on the 'coal question' although, in his general theory of utility published a few years later, he succeeded in ignoring this problem. What could Jevons have done? He had come upon the structural deficit inherent in the harmonic cosmos of the liberal market economy, which seemed to call for the intervention of a more far-sighted and rational agent than the market. Unlike some socialist theoreticians, however, he did not believe that such an agent really existed.

4. Nuclear or Solar Energy

The coal problem was a great challenge in economic theory formation. It appeared as if a resource barrier of the capitalist system was becoming apparent there, to which it simply could not respond optimally in terms of its own laws. The harmonic order of the self-regulating counterweights of price and value had suffered a blow in matters of energy that put the liberal principle of free pursuit of self-interest in doubt.

Of course, there was a series of attempts to get a theoretical grip on the problem (see Barnett and Morse 1963; Martinez-Alier 1987). A possible solution was treating depletable natural goods in a manner analogous to land by expressing relative scarcity as a rent in which expectations are discounted over time. (see Hotelling 1931; Solow 1974). If it were assumed that a rising scarcity rent stimulates technological progress or a shift to substituting raw materials, market forces would also signal a reasonable reaction in this case. But we must not forget that for classical economists technological progress only played a marginal role and was seen as an exogenous factor, not as an anticipated reaction to scarcity and price increase. However, if technological progress is understood as a reaction to market forces, it can become a major factor. As long as technological innovation and an escape to ever more common and more accessible alternative resources could take place more quickly than conventional resources were exhausted, there need not be a limit to growth.

In this process energy plays a key role. The transition from a more accessible but exhaustible resource to an alternative resource is often associated with growing energy inputs. This relationship is most clearly evident in the use of mineral raw materials. The quantity of the respective chemical elements of the earth roughly remains constant but these elements are distributed in different ways in the earth's crust. 'Consumption' of raw materials then means that certain elements are diffused so that their concentration will decline in the end. It can be anticipated that in the course of time ever more unfavourable raw material deposits that are more difficult to access, or more contaminated or in which the desired substance is contained in ever declining concentration must be exploited.

Therefore, the industrial process fundamentally operates as follows. It begins with the acquisition of materials available in relatively high concentrations such as ore deposits. In the next step these substances are further concentrated using energy, e. g. when metals are smelted. After that a diffusion of the materials caused by use sets in, for example when metals rust or are worn. In the final stage they will be widely distributed in stable chemical compounds over the surface of the earth and in the seas. From this final state there is hardly any recycling of substances possible, unless great quantities of cheap energy are available, which in the extreme case may permit recovering salts from the seas.

The same applies to energy raw materials in the narrow sense. There is a counter-tendency to growing scarcity and costliness in energy conserving techniques but no guarantee exists that technical efficiency will be improved faster than the energy requirements of raw material production will increase. All resource problems may in the end be represented as energy problems. Energy is the central resource of the industrial system. Inversely this means: if making energy available in superabundance succeeds, all other raw material and environmental problems are open to solutions.

> A technology that opens a virtually inexhaustible energy source would not only represent an input factor of basically unparalleled quantity that could directly secure a continuous growth of production and consumption but also would be *the one* solution to all other resource problems in the long term. Seen in economic terms, this would be a 'super-backstop-technology' that would not only temporarily remove the possibility of strangling economic development with rising rents but forever and for all times eliminate it (Meixner 1981, 45).

To this may merely be added that it is not only a continuation of economic growth and 'development' at stake, but also the long-term maintenance of the achieved level of material flows in the face of dwindling resources. Jevons had anticipated that there would be a historic reversal of industrialisation because the coal supply would run out. This fear seemingly lost its basis with the huge mineral oil deposits of the USA and the Middle East although his argument applies as much to them as to coal. The exhaustibility of energy resources remains a sword of Damocles hanging over the industrial system.

The hope of having found the 'super-backstop-technology', that final supply of resources eliminating all barriers to economic growth, has been associated with nuclear energy since the 1950s. Of, course, the customary reactor types relied on the use of a relatively small stock of uranium. But if breeder technology and finally nuclear fusion were mastered, resources so huge would become available that in the 'nuclear age' all scarcity would be overcome. Attitudes like those of Jevons would definitely be relegated to the past. A technical utopia would have come into reach that would leap across all growth boundaries.

However, nuclear energy has disadvantages that result from the nature of nuclear fission and are therefore unavoidable. With nuclear energy large quantities of radioactive radiation are released in a very small space. Also, radioactive waste is produced that is difficult to dispose because of the long time spans involved. A nuclear power plant is an extraordinarily complex system. Its undisrupted function rests upon a series of preconditions that obviously cannot be easily and permanently fulfilled. Thus, fundamental risks are associated with nuclear fission that may perhaps be minimised by technical measures, but have caused considerable public concern to the point of political resistance.

If we compare fossil energy with nuclear energy, the following is noticeable: fossil fuels possess almost ideal properties for humans. The energy density of coal is almost double that of firewood with otherwise similar properties. The use of coal and other fossil fuels like petrol and its derivatives like gasoline is largely unproblematic. Automobiles only explode in Hollywood movies. Gasoline possesses a far higher hydrogen content than coal, which is why it is lighter and its energy content per unit of weight is almost 50% higher. It is liquid like water and burns better than wood.

In contrast to nuclear energy the temperatures that are achieved when burning coal are usually close to what is required for commercial purposes. There is no lethal radiation and therefore no complex systems are required for handling these substances. To generate electricity, water is heated to drive turbines with steam that has a temperature of not much more than 100°C. In a coal power plant approximately the desired temperatures are generated. By contrast, a nuclear power plant will generate temperatures that are larger by orders of magnitude than what is required.

In contrast to hydrogen, fossil fuels do not have to be cooled or pressurised. In contrast to electricity, contact with them is harmless and neither the consumer nor they require protection. These properties were the reason why they were able to substitute wood and charcoal gradually without requiring extensive technological innovations and complicated systems from the start. The transition from the agrarian solar energy system to the fossil energy system occurred slowly and did not require a technological tour de force.

Technological utilisation of solar energy has been discussed for some time as an alternative to nuclear energy. Without doubt it does not possess the potential for risk associated with nuclear energy. However, the technical utilisation of solar energy also has disadvantages of a basic nature compared to fossil energy: solar energy essentially has low energy density. This applies as much to the sun's radiation as it does to the motion of wind and running water. If this energy is to be used, it must first be concentrated, which always involves a great effort in area, material and work. For example, technical collector systems that generate photovoltaic energy must be built, maintained, repaired and finally disposed of. If they are not to be subsidised with energy from other sources and their entire life cycle has to be maintained with solar energy alone, it is questionable if they can operate with a positive energy harvest factor. Even if the transformation of radiation into chemical energy is provided by plants, the labour of collecting the plants from the surface on which they grow is unavoidable. As we have seen, considerable transportation costs arise that in the end are energy costs and must be converted into the standard measure of the solar energy system, units of area.

Much the same applies to wind and water power. If one wishes to create reservoirs to produce hydroelectric energy, huge amounts of material must be moved for the construction of the dam, also areas must be flooded so they are withdrawn from other uses. By analogy, the same applies to the construction of wind power for which large, often high energy input materials like aluminium are used. Therefore, one may be sceptical with respect to the energy yield over their entire life cycle. The basic problem of the solar energy system is simply unavoidable: if energy of a low density is converted into a useful form, the inputs of area and material are enormous and fundamentally reduce the energy yield.

From the perspective of a technological and industrial solar energy system conceivable in the future, it should be considered that in the low temperature range the energy valency of fuels is essentially too high. If it is merely an issue of heating a room from 10° C to 20°C, a temperature of 400° C, as generated by combustion, is not required. The same effect can be achieved in a simpler manner with passive solar energy use. Here there is clearly room for the improvement of technical efficiency. The fundamental characteristics of the solar energy system also apply under technical and industrial conditions. The land area required to collect photosynthetically and to utilise solar energy determines in the end possible population density and the volume of material flows that can pass through the population.

Therefore, each of the three energy systems has its own disadvantages. Fossil energy is exhaustible and associated with CO_2 emissions into the atmosphere. Nuclear energy emits radioactive radiation and harbours the danger of uncontrollable failures. Solar energy requires large areas and large quantities of material so that the harvest is problematic. Other alternatives to these three are not in sight. Due to its existing properties, it may be anticipated that fossil energy still has a long future ahead of it.

Bibliography

Abel, Wilhelm 1955. *Die Wüstungen des ausgehenden Mittelalters.* Stuttgart.

Abel, Wilhelm 1972. *Massenarmut und Hungerkrisen im vorindustriellen Deutschland.* Göttingen.

Abel, Wilhelm 1978. *Agrarkrisen und Agrarkonjunktur.* Hamburg and Berlin.

Adams, Richard N. 1975. *Energy and Structure. A Theory of Social Power.* Austin.

Adams, Richard N. 1978. Man, Energy, and Anthropology: I can feel the heat, but where's the light? *American Anthropologist* 80, 297–309.

Agricola, Georg 1556. *De re metallica.* German edition, München 1977.

Agricola, Georg, 1612. *De natura fossilium.* Wittenberg.

Alberti 1766. Versuch über den Nutzen des Steinkohlen-Brandes in Stuben-Oefen. *Hannoverisches Magazin,* 10.1.1766, 42ff.

Alberti 1772. Versuchter Vorschlag, zur Ersparung des Holzes und Torfes sich statt deren Gebrauchs bey Heizung der Stubenöfen und des Küchenfeuers der Steinkohle zu bedienen. *Hannoverisches Magazin,* 24.2.1772, 242ff.

Alberti 1790. Über Steinkohlenbrand in Stubenöfen. *Hannoverisches Magazin,* 1.2.1790, 153ff.

Albion, R.G., 1926. *Forests and Sea Power.* Cambridge, Mass.

Allen, R., 1994. Agriculture during the industrial revolution. In: Floud, R. and McCloskey, D. (eds), *The Economic History of Britain since 1700,* Vol. 1, 96–122. Cambridge.

Allgemeine Deutsche Bibliothek. Berlin.

Allmann, Joachim 1989. *Der Wald in der frühen Neuzeit. Eine mentalitäts- und sozialgeschichtliche Untersuchung am Beispiel des pfälzer Raumes, 1500–1800.* Berlin.

Anmerkungen über den giftigen und tödlichen Dampf derer Schmiedekohlen. *Leipziger Sammlungen* Bd. 7, 1751, 231–35.

Arndt, E.M., 1815. Ein Wort über die Pflegung und Erhaltung der Forsten und der Bauern im Sinne einer höheren, d.h. menschlichen Gesetzgebung. In: Bartelmeß, A. *Wald. Umwelt des Menschen.* Freiburg and München, 1972.

Ashton, T.S., 1924. *Iron and Steel in the Industrial Revolution.* Manchester.

Aström, S.E., 1970. English Timber Imports from Northern Europe in the 18th Century. *Scandinavian Economic History Review* 18, 12–32.

Aström, S.E., 1975. Technology and Timber Exports from the Gulf of Finland, 1661–1740. *Scandinavian Economic History Review* 23, 1–14.

Aström, S.E., 1978. Foreign Trade and Forest Use in Northeastern Europe, 1660–1860. *Natural Resources* 1978, 43–64.

Baechler, J.; Hall, J.A. and Mann, M. (eds) 1988. *Europe and the Rise of Capitalism.* Oxford.

Bairoch, Paul 1978. Agriculture and the Industrial Revolution, 1700–1914. In: *Fontana Economic History of Europe*, Vol. 3, 452–506. Glasgow.

Bairoch, Paul 1993. *Economics and World History. Myths and Paradoxes.* New York.

Bakewell, R. 1828. *An Introduction to Geology.* London.

Barnett, H.J. and Morse, C., 1963. *Scarcity and Growth. The Economics of Natural Resource Availability.* Baltimore.

Bayerl, G., 1989. *Wind- und Wasserkraft. Die Nutzung regenerierbarer Energiequellen in der Geschichte.* Düsseldorf.

Beantwortung der Preisfrage: Wie dem einreißenden Holzmangel vorzubeugen sei. Erfurt 1764.

Beaumont, C., 1789. *Treatise on the Coal Trade.* London.

Bemerkungen eines Reisenden durch Deutschland, Frankreich, England und Holland. Altenburg 1775.

Bernhardt, A., 1872, 1874, 1875. *Geschichte des Waldeigentums, der Waldwirtschaft und Forstwissenschaft in Deutschland.* 3 Vols. Berlin.

Blasius, D., 1975. Eigentum und Strafe. *Historische Zeitschrift* 220, 79–129.

Blickle. P., 1986. Wem gehört der Wald? Konflikte zwischen Bauern und Obrigkeiten um Nutzungs- und Eigentumsansprüche. *Zschr. für württembergische Landesgeschichte* 45, 157–178.

Bloch, M., 1977. Antritt und Siegeszug der Wassermühle. In: Honegger, C. (ed.), *Schrift und Materie der Geschichte,* 171–97. Frankfurt am Main.

Bloss, O., 1977. *Die älteren Glashütten in Südniedersachsen.* Hildesheim.

Bogucka, M., 1978. North European Commerce as a Solution to Resource Shortage in the 16th–18th Centuries. In: A. Maczak and W.N. Parker (eds) *Natural Resources in European History,* 9–42. Washington.

Bogucka, M., 1980. The Role of Baltic Trade in European Development from the 16th to the 18th Centuries. *Journal of European Economic History* 9, 5ff.

Borgstrom, G., 1972. *The Hungry Planet.* New York.

Bornemann, C.E., 1776. *Versuch einer systematischen Abhandlung von den Kohlen.* Göttingen.

Boserup, E., 1965. *The Conditions of Agricultural Growth. The Economics of Agrarian Change under Population Pressure.* London.

Boyden, S., 1987. *Western Civilization in Biological Perspective.* Oxford.

Braudel, F. 1974. *Capitalism and Material Life, 1400–1800.* Glasgow.

Brockhaus Conversationslexikon. 12th edition. Leipzig 1875ff.

Buckland, W., 1836. *Geology and Mineralogy* (Bridgewater Treatise, 6). London.

Bülow, G., 1962. Die Sudwälder von Reichenhall. In: *Mitteilungen aus der Staatsforstverwaltung Bayerns* 33. München.

Bünting, J.P., 1693. *Sylva Subterranea, oder Vortreffliche Nutzbarkeit des Unterirdischen Waldes der Steinkohlen.* Halle.

Burgsdorf, F.A.L.v., 1790. Abhandlung über die Vortheile vom ungesäumten, ausgedehnten Anbau einiger in den Königl. Preußischen Staaten noch ungewöhnlichen Holzarten. Vortrag vor der Königl. Akademie der Wissenschaften zu Berlin am 14.1.1790. In: W.G.v. Moser, *Forst-Archiv,* 8, 265–93. Ulm.

Buxton, N.K., 1978. *The Economic Development of the British Coal Industry.* London.

Cardwell, D.S.L., 1963. *Steam Power in the 18th Century.* London.

Cardwell, D.S.L., 1965. Power Technologies and the Advance of Science, 1700–1825. *Technology and Culture* 6, 188–207.

Chambers, J.D. and Mingay, G.E., 1966. *The Agricultural Revolution, 1750–1880.* London.

Childe, G., 1942. *What Happened in History?* Harmondsworth 1978.

Cipolla, C.M., 1962. *The Economic History of World Population.* Harmondsworth.

Clow, A. and Clow, N., 1952. *The Chemical Revolution.* London.

Clow, A. and Clow, N., 1956. The Timber Famine and the Development of Technology. *Annals of Science* 12, 85–102.

Cohen, M.N., 1977. *The Food Crisis in Prehistory. Overpopulation and the Origins of Agriculture.* New Haven and London.

Cohen, M.N., 1989. *Health and the Rise of Civilization.* New Haven.

Coleman, D.C., 1977. *The Economy of England, 1450–1750.* London.

Conybeare, W.D. and Phillips, W., 1822. *Outlines of Geology.* London.

Cottrell, F., 1955. *Energy and Society.* New York.

Crew, D., 1979. *Town in the Ruhr. A Social History of Bochum, 1860–1914.* New York.

Crone, P., 1989. *Pre-industrial Societies.* Oxford.

Crosby, Alfred W. 1972. *The Columbian Exchange. Biological and Cultural Consequences of 1492.* Westport.

Crosby, Alfred W. 1986. *Ecological Imperialism. The Biological Expansion of Europe, 900–1900.* Cambridge.

Cundy, N.W., 1833. *Inland Transit. The Practicability, Utility, and Benefit of Railroads.* Report of a Select Committee of the House of Commons on Steam Carriages. London.

Dahlstein, G., 1797. *Anleitung zum gemeinnützigen Gebrauch von Steinkohle.* Wien.

Darby, H.C., 1950. Domesday Woodland. *Economic History Review* 3, 21–43.

Darby, H.C., 1951. The Clearing of the English Woodlands. *Geography* 36, 71–83.

Darby, H.C., 1956. The Clearing of the Woodland in Europe. In: Thomas, W.L. (ed.), *Man's Role in Changing the Face of the Earth.*, 183–216. Chicago.

Davis, R., 1962. *The Rise of the English Shipping Industry.* London.

de Vries, J. and van der Woude, A. 1997. *The First Modern Economy. Success, Failure, and Perseverance of the Dutch Economy.* Cambridge.

Deane, P., 1965. *The First Industrial Revolution.* Cambridge.

Donovan, S.K., 1989. *Introduction to Mass Extinctions. Processes and Evidence.* New York.

Dyer, A.D., 1976. Wood and Coal: A Change of Fuel. *History Today* 26, 598–607.

Dyos, H.J. and Aldcroft, D.H., 1974. *British Transport. An Economic Survey from the 17th Century to the 20th.* Harmondsworth.

Eckhardt, H.W., 1976. *Herrschaftliche Jagd, bäuerliche Not und bürgerliche Kritik.* Göttingen.

Eichmann, O.L.v., 1783. *Über die Vorzüge der Feuerung mit Steinkohlen.* Halle.

Eisenstadt, S.N. (ed.), 1967. *The Decline of Empires.* Englewood Cliffs.

Endres, M., 1927. Art. Forsten. In: *Handwörterbuch der Staatswissenschaft*, Bd. 4, 248–55. Jena.

Ertle, G.J.M., 1957. Von der „Holzspahrkunst" im 18. Jahrhundert. *Holz-Zentralblatt* 24.8.1957, 1245–47.

Etwas über den Gebrauch und Verbrauch von Steinkohlen in England. *Hannoverisches Magazin.* 26.1.1789, 122ff.

Etwas über die Steinkohlenfeuerung in Schlesien und England. *Hannoverisches Magazin.* 25.12.1789, 1643ff.

Evelyn, J. 1661. *Fumifugium, or, The Inconveniencie of the Aer and Smoak of London dissipated.* London.

Evelyn, J., 1664. *Sylva, or a Discourse of Forest-Trees, and the Propagation of Timber.* London.

Ewald, P., 1994. *The Evolution of Infectious Disease.* Oxford.

Faber, A., 1940a. Rauchplage und Wärmetheorie im 18. Jahrhundert. *Haustechnische Rundschau* 45.

Faber, A., 1940b. Kohle statt Holz zur Zeit Friedrichs des Großen. *Organ für das Schornsteinfegerwesen* 67, Heft 5–7, 49ff.

Faber, A., 1941. Geschichtliches zur Umstellung von Holzbrand auf Kohlefeuerung in Haushalt und Gewerbe. In: *Neue Deutsche Töpferzeitung.* Beilage Kachelofen, 33.

Faber, A., 1950. 1000 Jahre Werdegang von Herd und Ofen. *Deutsches Museum. Abhandlungen und Berichte* 18, Heft 3.

Faber, A., 1957. *Entwicklungsstufen der häuslichen Heizung.* München.

Fechner, H., 1900/1902. Geschichte des schlesischen Berg- und Hüttenwesens in der Zeit Friedrichs des Großen, Friedrich Wilhelms II. und Friedrich Wilhelms III., 1741–1806. *Zeitschrift für Berg- Hütten- und Salinenwesen*, 48, 1900, 279–401; 49, 1901, 1–86, 243–88, 383–446, 487–568; 50, 1902, 140–228, 243–310, 415–505, 691–796.

Fischer-Kowalski, M. (ed), 1997. *Gesellschaftlicher Stoffwechsel und Kolonisierung von Natur.* Amsterdam.

Fischer-Kowalski, Marina and Weisz, Helga 1999. Society as Hybrid between Material and Symbolic Realms. Toward a theoretical framework of society-nature interaction. *Advances in Human Ecology* 8, 215–215.

Flinn, M.W., 1958. The Growth of the English Iron Industry, 1660–1760. *Economic History Review* 11, 144–53.

Flinn, M.W., 1959a. Abraham Darby and the Coke Smelting Process. *Economica*, N.S. 26, 154–59.

Flinn, M.W., 1959b. Timber and the Advance of Technology. A Reconsideration. *Annals of Science* 15, 109–120.

Flinn, M.W., 1967. Consommation du bois et developpement sidérurgique en Angleterre. *Actes du Colloque sur la Forêt. Cahiers d'Etudes Comtoires* 12, 107–122.

Flinn, M.W., 1978. Technical Change as an Escape from Resource Scarcity. England in the 17th and 18th Centuries. *Natural Resources* 1978, 139ff.

Fordyce, W., 1860. *History of Coal, Coke, and Coal Mining.* Newcastle.

Frank, A.G., 1998. *ReOrient. Global Economy in the Asian Age.* Berkeley.

Franz, F.C., 1795. *Beantwortung der Frage: Wie dem Holzmangel vorzubeugen sei.* Leipzig.

Galloway, R.L., 1882. *History of Coal Mining in Great Britain.* London.

Ganzenmüller, W., 1956. *Beiträge zur Geschichte der Technologie und Alchemie.* Weinheim.

Gates, D.M., 1971. The Flow of Energy in the Biosphere. In: Scientific American Books (no editor), *Energy and Power*, 43–51. San Francisco.

Gellner, E., 1988. *Plough, Sword and Book.* London.

Georgescu-Roegen, Nicholas 1971. *The Entropy Law and the Economic Process.* Cambridge.

Gimpel, J., 1980. *Die industrielle Revolution des Mittelalters.* Zürich and München.

Glacken, Clarence J. 1967. *Traces on the Rhodian Shore. Nature and Culture in Western Thought from Ancient Times to the End of the 18th Century.* Berkeley.

Gleitsmann, R.J., 1980. Rohstoffmangel und Lösungsstrategien. Das Problem vorindustrieller Holzknappheit. *Technologie und Politik* 16, 104–54.

Gleitsmann, R.J., 1982. Aspekte der Ressourcenproblematik in historischer Sicht. *Scripta Mercaturae* 16, Heft 1.

Goodman, Godfrey 1616. *The Fall of Man, or the Corruption of Nature.* London.

Goudsblom, Johan 1992. *Fire and Civilization.* London.

Grabas, Margrit 1995. Krisenbewältigung oder Modernisierungsblockade? Die Rolle des Staates bei der Überwindung des „Holzmangels" zu Beginn der Industriellen Revolution in Deutschland. *Jahrbuch für Verwaltungsgeschichte* 7, 43–.

Grayson, D.K., 1980. Vicissitudes and Overkill. The Development of Explanations of Pleistocene Extinctions. *Advances in Archaeological Method and Theory* 3, 357–403.

Grimm, J., 1957. *Weisthümer*, Bd. 4. Darmstadt.

Grimm, J. and Grimm, W. (eds), 1854–. *Deutsches Wörterbuch.* Leipzig.

Groh, D., 1992. Strategien, Zeit und Ressourcen. Risikominimierung, Unterproduktivität und Mußepräferenz – die zentralen Kategorien von Subsistenzökonomien. In: *Anthropologische Dimensionen der Geschichte*, 54–113. Frankfurt a.M.

Guilmartin, J. 1974. *Gunpowder and Galleys*. Cambridge.

Hahnemann, S., 1787. *Abhandlung über die Vorurtheile gegen die Steinkohlenfeuerung*. Dresden.

Hall, J.A., 1985. *Powers and Liberties. The Causes und Consequences of the Rise of the West*. Oxford.

Hammersley, G., 1957. The Crown Woods and Their Exploitation in the 16th and 17th Centuries. *Bulletin of the Institute of Historical Research* 30, 136–161.

Hammersley, G., 1973. The Charcoal Iron Industry and its Fuel, 1540–1750. *Economic History Review* 26, 593–613.

Hammersley, G., 1979. Did it Fall or Was it Pushed? The Foleys and the End of the Charcoal Iron Industry in the 18th Century. In: Smout, T.C. (ed.), *The Search for Wealth and Stability*, 67–90. London.

Hardin, G., 1968. The Tragedy of the Commons. *Science* 162, 1243–48.

Harnisch, Hartmut 1997. Die Energiekrise des 18. Jahrhunderts als Problem der preußischen Staatswirtschaft. In: H.J. Gerhard (ed.), *Struktur und Dimension. Festschrift für K.H. Kaufhold*, Bd. 1, 589–10. Stuttgart.

Harris, Marvin 1978. *Cannibals and Kings. The Origins of Culture*. New York.

Harris, Marvin 1979. *Cultural Materialism. The Struggle for a Science of Culture*. New York.

Harris, Marvin and Ross, E.B., 1987. *Death, Sex, and Fertility. Population Regulation in Preiindustrial and Developing Societies*. New York.

Harris, V. 1949. *All Coherence Gone*. Chicago.

Hartmann, K.F.A., 1834. *Vollständige Brennmaterialkunde* (Neuer Schauplatz der Künste und Handwerke, 112). Weimar and Leipzig.

Hartwell, Robert 1962. A Revolution in the Chinese Iron and Coal Industries During the Northern Sung, 960–1126 A.D. *Journal of Asian Studies* 21, 153–62.

Hartwell, Robert 1967. A Cycle of Economic Change in Imperial China. Coal and Iron in Northwest China, 750–1350. *Journal of the Social and Economic History of the Orient* 10, 102–59.

Haßlacher, A., 1884. *Der Steinkohlenbergbau des preußischen Staates in der Umgebung von Saarbrücken*. Bd. 2: Geschichtliche Entwicklung. Berlin.

Hausmann, W.S., 1977. Size and Profitability of English Colliers in the 18th Century. *Business History Review* 51, 460–73.

Hausrath, H., 1907. *Der deutsche Wald*. Leipzig.

Henning, F.W., 1978/9. *Landwirtschaft und ländliche Gesellschaft in Deutschland*. Bd. 1, 800–1750; Bd.2, 1750–1976. Paderborn.

Herlihy, David 1997. *The Black Death and the Transformation of the West*. Cambridge.

Hills, R.L., 1970. *Power in the Industrial Revolution*. New York.

Historical Statistics of the United States. Washington 1976

Hoffmann, F., 1695. *Propempticon inaugurale de vapore carbonum fossilium innoxio.* Halle.

Hoffmann, F., 1708. *Kurtze doch gründliche Beschreibung des Saltz-Wercks in Halle.* Halle.

Hoffmann, F., 1716. *Gründliches Bedenken und physicalische Anmerkungen von dem tödtlichen Dampff der Holtz-Kohlen.* Halle.

Holland, J., 1841 (2nd edn). *A History and Description of Fossil Fuel, the Collieries and Coal Trade of Great Britain.* London.

Hotelling, H., 1931. The Economics of Exhaustible Resources. *Journal of Political Economy* 39, 137–75.

House of Commons, 1830. *Report of the Select Committee on the State of the Coal Trade.* London.

House of Commons, 1871. *Report of the Commissioners appointed to Inquire into several matters relating to Coal.* London.

House of Lords, 1829. *Report from the Select Committee of the House of Lords. The State of the Coal Trade.* London.

Huber, J., 1717/1762. *Curieuses und Reales Natur-Kunst-Berg-Gewerck- und Handlungs-Lexicon.* Leipzig.

Huberti, J.L., 1765. *Abhandlung von dem allgemeinen Holzmangel und von den Mitteln, solchem zu steuern.* Frankfurt am Main.

Hue, O., 1910. *Die Bergarbeiter.* Vol. 1. Stuttgart.

Hughes, J.D., 1975. *Ecology in Ancient Civilizations.* Albuquerque.

Hughes, J.D., 1996. *Pan's Travail. Environmental Problems of the Ancient Greeks and Romans.* Baltimore.

Hyde, C.K., 1973. The Adoption of Coke-Smelting by the British Iron Industry, 1709–1790. *Explorations in Economic History* 10, 397–418.

Hyde, C.K., 1977. *Technological Change and the British Iron Industry, 1700–1870.* Princeton.

Jachtmann, H., 1794. *Abhandlungen von Anlegung der Brau- und Branntweinbrennerey- und Malzdarrenfeuerungen zum ersparenden Holz-, Steinkohlen- und Torfbrande.* Berlin.

Jäger, H., 1994. *Einführung in die Umweltgeschichte.* Darmstadt.

Jars, G., 1774–1781. *Voyages métallurgiques,* 3 Bde. Lyon.

Jevons, W.S., 1865. *The Coal Question.* London.

Jevons, W.S., 1906. *The Coal Question.* 3rd edition, ed. A.W. Flux. London.

Johann, E., 1968. *Geschichte der Waldnutzung in Kärnten unter dem Einfluß der Berg-, Hütten- und Hammerwerke.* Klagenfurt.

Jones, E.L., 1987. *The European Miracle. Environments, Economics and Geopolitics in the History of Europe and Asia,* 2nd edition, Cambridge.

Justi, J.H.G.v., 1761. Von der Aufmerksamkeit eines Cameralisten auf die Waldungen und den Holzanbau. In: *Gesammelte Politische und Finanzschriften.* Vol. 1, 439–64. Kopenhagen and Leipzig.

Kahlert, W., 1955. Die Wärmewirtschaft mittelalterlicher Glasschmelzöfen. *Glastechnische Berichte* 28, 483–485.

Kaufhold, K.H., 1976. *Die Metallgewerbe der Grafschaft Mark im 18. und frühen 19. Jahrhundert.* Dortmund.

Keegan, J., 1993. *A History of Warfare.* London.

Keeley, Lawrence H., 1996. *War before Civilization.* New York.

Kemp, W.B., 1971. The Flow of Energy in a Hunting Society. In: Scientific American Books (no editor), *Energy and Power,* 55–65. San Francisco.

Kent, H.S.K., 1955/6. The Anglo-Norwegian Timber Trade in the 18th Century. *Economic History Review* 8, 62–74.

Kerridge, E., 1967. *The Agricultural Revolution.* London.

Kerridge, E., 1977. The Coal Industry in Tudor and Stuart England. *Economic History Review* 30, 340–342.

Kessler, F., 1618. *Holz-spahr-Kunst.* Frankfurt am Main.

Kiple, Kenneth F. and Beck, Stephen V. (eds) 1997. *Biological Consequences of the European Expansion, 1450–1800.* Aldershot.

Kjaergaard, T., 1994. *The Danish Revolution, 1500–1800. An Ecohistorical Interpretation.* Cambridge.

Kjaerheim, S., 1957. Norwegian Timber Exports in the 18th Century. *Scandinavian Economic History Review* 5, 188–202.

Klein, R., 1992. The Impact of Early People on the Environment. The Case of Large Mammal Extinctions. In: Jacobsen, J.E. and Firor, J. (eds.), *Human Impact on the Environment,* 13–34. Boulder.

Komlos, John and Baten, J. (eds) 1998. *Studies in the Biological Standard of Living in Comparative Perspective.* Stuttgart.

Köstler, J., 1934. *Geschichte des Waldes in Altbayern.* München.

Krüger, J.G., 1741. *Gedancken von den Stein-Kohlen.* Halle.

Krünitz, J.G. (ed.), 1773–1858. *Oeconomische Encyclopädie oder allgemeines System der Land-, Haus- und Staatswirthschaft.* 242 vols. Berlin.

Kürten, W.v., 1972. Die Entwicklung des Steinkohlenreviers an der unteren Ruhr bis zur Mitte des 19. Jahrhunderts. *Beiträge zur Heimatkunde der Stadt Schwelm und ihrer Umgebung.* N.F. 22, 42–69.

Küster, Hansjörg 1995. *Geschichte der Landschaft in Mitteleuropa.* München.

Lagus, H., 1908. *Über den Holzexport und die Wälder Finnlands.* Dissertation, Heidelberg.

Landes, D.S., 1969. *The Unbound Prometheus.* Cambridge.

Landes, D.S., 1998. *The Wealth and Poverty of Nations.* New York.

Le Roy Ladurie, E., 1971. *Times of Feast, Times of Famine. A History of Climate Since the Year 1000.* Garden City.

Le Roy Ladurie, E. 1973. Un concept: L'unification microbienne du monde (XIV–XVIIIe siècles). *Schweizerische Zeitschrift für Geschichte* 23, 627–96.

Leach, G., 1976. *Energy and Food Production.* Guildford.

Lee, R.B., 1968. What Hunters Do for a Living or How to Make Out on Scarce Resources. In: Lee, R.B and DeVore, I. (eds), *Man the Hunter,* 30–43. Chicago.

Lee, R.B., 1979. *The !Kung San.* Cambridge.

Lehmann, J.C., 1723. *Ars lucrandi ligni, das ist Universal-Holzsparkunst.* Leipzig.

Lemnius, Levinus, 1666. *De miraculis occultis naturae.* Leyden.

Leutmann, M.J.G., 1720. *Vulcanus Famulans oder Sonderbahre Feuer-Nutzung.* Wittenberg.

Lewis, M.J.T., 1970. *Early Wooden Railways.* London.

Lichtenberg, G.C., 1776. Sudelbücher. In: *Schriften und Briefe.* Vol. 1, München 1968.

Lindemann, C.F.H., 1784. *Reisebemerkungen über einen Theil von Italien, Frankreich und Engelland.* Celle.

Lindsay, J.J., 1975. Charcoal Iron Smelting and its Fuel Supply: The Example of Lorn Furnace, Argyllshire, 1753–1876. *Journal of Historical Geography* 1, 283–298.

Livi-Bacci, M., 1997. *A Concise History of World Population.* 2nd edition. Oxford.

Loomis, R.S., 1978. Ecological dimensions of Medieval Agrarian Systems. *Agricultural History* 52, 478–483.

Lovins, Amory 1977. *Soft Energy Paths – Towards a Durable Peace.* New York.

Lüning, J., 1988. Frühe Bauern in Mitteleuropa im 6. und 5. Jahrtausend v. Chr. In: *Jahrbuch des Römisch-Germanischen Zentralmuseums,* 35. Mainz.

Lutz, J., 1941. Die ehemaligen Eisenhämmer und Hüttenwerke der nördlichen Oberpfalz. In: *Mitteilungen aus Forstwirtschaft und Forstwissenschaft* 12. Hannover.

MacNab, H.G., 1792. *Treatise on the Coal Trade.* London.

Madihn, J.J., 1772. Von den Steinkohlen. *Hannoverisches Magazin,* 18.9.1772, 1190ff.

Mager, F., 1960. *Der Wald in Altpreußen als Wirtschaftsraum.* Köln.

Malthus, T.R., 1826. *An Essay on the Principle of Population,* 6th edn. London 1973.

Mantel, K., 1975. *Entwicklungslinien der Forstwirtschaft vom Mittelalter bis zum 19. Jahrhundert.* Berlin and New York.

Marek, Daniel 1994. Der Weg zum fossilen Energiesystem. Ressourcengeschichte der Kohle am Beispiel der Schweiz 1850–1910. In: W. Abelshauser (ed.), *Umweltgeschichte.* Göttingen, 57–75.

Martin, P.S. and Wright, H.E. (eds), 1967. *Pleistocene Extinctions. The Search for a Cause.* New Haven.

Martinez-Alier, J., 1987. *Ecological Economics. Energy, Environment and Society.* Oxford.

Matossian, Mary K. 1986. Did Mycotoxins Play a Role in Bubonic Plague Epidemics? *Perspectives in Biology and Medicine* 29, 244–56.

Matschoß, C., 1908. *Die Entwicklung der Dampfmaschine,* 2 Vols. Berlin.

McCulloch, J.R., 1830. *Observations on the Duty on Sea-Borne Coal.* London.

McCulloch, J.R., 1832. *Dictionary of Commerce*. London.

McNeill, Willaim H., 1980. *The Human Condition. An Ecological and Historical View*. Princeton.

McNeill, William 1976. *Plagues and Peoples*. Garden City.

Medicus, F.K., 1768. *Von dem Bau auf Steinkohlen*. Mannheim.

Meisner, C.H., 1801. *Handbuch zur Holzersparung; oder: Anleitung, wie man sowohl Torf als auch Steinkohlen entdecken könne, und wie die Öfen eingerichtet seyn müssen, um diese Brennmittel mit weit mehrern Nutzen, als zeither geschehen, zur Feuerung anwenden zu können*. Leipzig.

Meixner, H., 1983. Die ökonomische Logik der Kernenergie. *Jahrbuch für Sozialwissenschaft* 34, 59–93.

Melosi, Martin V. 1982. Energy transitions in the 19th century economy. In: Daniels, G.H. and Rose, M.H. (eds), *Energy and Transport*, 55–67. Beverly Hills.

Merchant, C. 1980. *The Death of Nature. Women, Ecology, and the Scientific Revolution*. San Francisco.

Metallgesellschaft 1992. *Metallstatistik*, vol. 79. Frankfurt am M.

Mill, J.S., 1848. *Principles of Political Economy*. London.

Mitscherlich, G., 1963. *Zustand, Wachstum und Nutzung des Waldes im Wandel der Zeit*. Freiburg.

Möller, G.F., 1750. Ohnmaßgebliche Vorschläge, wie eine ohnfehlbare Holz-Erspahrung bey Erheizung der Wohn-Zimmer zu erhalten. In: *Oeconomische Nachrichten*, Vol. 1, 647–87. Leipzig.

Mols, R., 1974. Population in Europe, 1500–1700. In: *The Fontana Economic History of Europe*, Vol. 2, 15–82. Glasgow.

Montanari, Massimo 1993. *Der Hunger und der Überfluß. Kulturgeschichte der Ernährung in Europa*. München.

Morand, M., 1777. Memoires sur les feux de houille, ou charbon de terre. In: *L'art d'exploiter les mines de charbon de terre*, Vol. 2. Paris.

Morton, G.R. 1966. The Early Coke Era. *Bulletin of the Historical Metallurgy Group* 6, 49–60.

Möser, J., o.J. 1945. Für die warmen Stuben der Landleute. In: *Sämtliche Werke*, Vol. 5, 241–3. Oldenburg and Berlin.

Müller-Herold, U. and Sieferle, R.P., 1997. Surplus and Survival. Risk, Ruin, and Luxury in the Evolution of Early Forms of Subsistence. *Advances in Human Ecology* 6, 201–220.

Multhauf, R.P., 1978. *Neptune's Gift*. London.

Musson, A.E. and Robinson, E., 1969. *Science and Technology in the Industrial Revolution*. Manchester.

Needham, J., 1969. Science and Society in East and West. In: *The Grand Titration*, 190–217. London.

Needham, J., 1971. *Science and Civilization in China*, Vol. 4. London and New York.

Nef, J.U., 1932. *The Rise of the British Coal Industry.* London and Edinburgh.

Nef, J.U., 1934/35. The Progress of Technology and the Growth of Large-Scale Industry in Great Britain, 1540–1640. *Economic History Review* 5, 3–24.

Nef, J.U., 1957. Coal Mining and Utilization. In: Singer, C. (ed.), *A History of Technology,* vol. 3, 72–88. Oxford.

Nef, J.U., 1958. *Cultural Foundations of Industrial Civilization.* Cambridge.

Nef, J.U., 1977. An Early Energy Crisis and its Consequences. *Scientific American* 237, 5, 140–152.

Nourse, T., 1700. *Campania Foelix, or, a Discourse of the Benefits and Improvements of Husbandry.* Annex: Of the Fuel of London, 345ff. London.

Odum, E.P./Reichholf, J., 1980. *Ökologie.* München.

Odum, Eugene P. 1975. *Ecology. The Link Between the Natural and Social Sciences.* New York.

Ohler, N., 1986. *Reisen im Mittelalter.* München and Zürich.

Ostrom, E., 1990. *Governing the Commons. The Evolution of Institutions for Collective Action.* Cambridge.

Ostwald, W., 1909. *Energetische Grundlagen der Kulturwissenschaft.* Leipzig.

Ostwald, W., 1912. *Der energetische Imperativ.* Leipzig.

Pfeiffer, J.F.v., 1775. *Geschichte der Steinkohlen und des Torfs.* Mannheim.

Pfeiffer, J.F.v., 1781. *Grundriß der Forstwirtschaft.* Mannheim.

Pfeil, W., 1816. *Über die Ursachen des schlechten Zustandes der Forsten und die allein möglichen Mittel, ihn zu verbessern, mit besonderer Rücksicht auf die preußischen Staaten.* Züllichau and Freistadt.

Pfeil, W., 1832. Holzmangel. In: Ersch, J.S. and Gruber, J.G., (eds) *Allgemeine Encyclopaedie der Wissenschaften und Künste,* sect. 2, vol. 9, 165–7. Leipzig.

Pfeil, W., 1839. *Forstgeschichte Preußens bis 1806.* Leipzig.

Pfister, Christian 1999. *Wetternachhersage. 500 Jahre Klimavariationen und Naturkatastrophen, 1496–1995.* Bern.

Pfister, J.G. v., 1775. *Geschichte der Steinkohlen und des Torfs.* Mannheim.

Polanyi, K., 1978. *The Great Transformation.* Frankfurt am Main.

Pollard, S., 1980. A New Estimate of British Coal Production, 1750–1850. *Economic History Review* 33, 212–35.

Postlethwayt, M., 1747. *Considerations on the Making of Bar Iron with Pit or Sea Coal Fire.* London.

Radkau, J. 1983. Holzverknappung und Krisenbewußtsein im 18. Jahrhundert. *Geschichte und Gesellschaft* 9, 513–43.

Radkau, J. 1986. Zur angeblichen Energiekrise des 18. Jahrhunderts. Revisionistische Betrachtungen über die 'Holznot'. *Vierteljahrschrift für Sozial- und Wirtschaftsgeschichte* 73, 1–37.

Radkau, J. 1997. Das Rätsel der städtischen Brennholzversorgung im „hölzernen Zeitalter". In: Schott, Dieter (ed.), *Energie und Stadt in Europa.* Stuttgart, 43–75.

Rappaport, R.A. 1968. *Pigs for the Ancestors. Ritual in the Ecology of a New Guinea People.* New Haven.

Rappaport, R.A., 1971. The Flow of Energy in an Agricultural Society. In: Scientific American Books (no editor), *Energy and Power,* 69–80. San Francisco.

Reynolds, J., 1970. *Windmills and Watermills.* London.

Ricardo, D., 1821. On the Principles of Political Economy and Taxation. Works and Correspondence, Vol. 1. Cambridge 1962.

Robinson, E.H., 1974. The Early Diffusion of Steam Power. *Journal of Economic History* 34, 91–107.

Rolt, L.T.C., 1963. *Thomas Newcomen. The Prehistory of the Steam Engine.* London.

Rössing, Adelbert 1901. *Geschichte der Metalle.* Berlin.

Rubner, H., 1967. *Forstgeschichte im Zeitalter der industriellen Revolution.* Berlin.

Russell, J.C., 1972. Population in Europe, 500–1500. In: *The Fontana Economic History of Europe,* Vol. 1, 25–70. Glasgow.

Ryan, C.J. 1979. Energy and the structure of social systems: A Theory of Social Evolution. Paper presented at the 1979 annual meeting of the American Association for the Advancement of Science in Houston, Texas.

Sahlins, M, 1974. *Stone Age Economics.* London.

Sanderson, Stephen K. 1995. *Social Transformations. A General Theory of Historical Development.* Oxford.

Scheidt, K.A (undated). *Kurze Betrachtung über einige Ursachen des allgemein werdenden Holzmangels in Deutschland und über die Mittel, demselben abzuhelfen.* München.

Scheidt, K.A., 1768. Versuch einer practischen Anleitung, Steinkohlenlager in ihren gebürgen aufzusuchen und dieselben zu bearbeiten. *Medicus* 1768, 161ff.

Schmid, G.V., 1839. *Handbuch aller seit 1560 bis auf die neueste Zeit erschienenen Forst- und Jagdgesetze des Königreichs Sachsen.* Vol. 1. Meißen.

Schmieding, F., 1991. *Der Energieaufwand der technischen Schwefelsäureerzeugung. Ein Beitrag zur Energiegeschichte der chemischen Industrie.* Dissertation, Frankfurt am Main.

Schroth, K., 1912. *Geschichte der Verkehrs- und Absatzverhältnisse beim oberschlesischen Steinkohlenbergbau in den ersten 100 Jahren seiner Entwicklung, 1748–1845.* Dissertation, Breslau.

Schubert, H.R., 1957. *History of the British Iron and Steel Industry from 450 to A.D. 1775.* London.

Schulze, C.F., 1764. *Zufällige Gedanken über den Nutzen der Steinkohlen und des Torfes.* Friedrichsstadt.

Schwappach, A., 1886/1888. *Handbuch der Forst-und Jagdgeschichte.* 2 vols. Berlin.

Selter, Bernward 1995. *Waldnutzung und ländliche Gesellschaft. Landwirtschaftlicher „Nährwald" und neue Holzökonomie im Sauerland des 18. und 19. Jahrhunderts.* Paderborn.

Serlo, A., 1869. *Beiträge zur Geschichte des schlesischen Bergbaus in den letzten 100 Jahren.* Breslau.

Sharp, L., 1975. Timber, Science, and Economic Reform in the 17th Century. *Forestry* 48, 51–86.

Sieferle, R.P., 1997. Kulturelle Evolution des Gesellschaft-Natur-Verhältnisses In: Fischer-Kowalski, M. (ed.), 1997. *Gesellschaftlicher Stoffwechsel und Kolonisierung von Natur.* Amsterdam, 37–53.

Sieglerschmidt, Jörn 1999. Wandlungen des Energieeinsatzes in Mitteleuropa in der Frühneuzeit. *Tübinger Geographische Studien* 125, 47–89.

Smil, Vaclav, 1991. *General Energetics. Energy in the Biosphere and Civilization.* New York.

Smil, Vaclav 1994. *Energy in World History.* Boulder.

Smith, A., 1776. *An Inquiry into the Nature and Causes of the Wealth of Nations.* The Glasgow Edition of the Works and Correspondence of Adam Smith. Oxford 1976.

Smith, B.D., 1995. *The Emergence of Agriculture.* New York.

Smith, C.T., 1978. *An Historical Geography of Western Europe before 1800.* London and New York.

Smith, R., 1961. *Sea-Coal for London.* London.

Soddy, Frederick 1921. *Cartesian Economics. The Bearing of Physical Science upon State Stewardship.* London.

Soederlund, E.F.S., 1960. The Impact of the British Industrial Revolution on the Swedish Iron Industry. In: Pressnell, L.S. (ed.), *Studies in the Industrial Revolution.* London. 52–65.

Solow, R.M., 1974. The Economics of Resources and the Resources of Economics. *American Economic Review* 64, 2, 1–14.

Sombart, Werner 1902 and 1919. *Der moderne Kapitalismus.* Leipzig.

Sombart, Werner 1913. *Die deutsche Volkswirtschaft im 19. Jahrhundert.* Berlin.

Sprandel, R., 1968. Das *Eisengewerbe im Mittelalter.* Stuttgart.

Stannard, David E. 1992. *American Holocaust. Columbus and the Conquest of the New World.* New York.

Steward, Julian 1955. *Theory of Culture Change.* Urbana.

Stöckhardt, A. 1850. Über die Einwirkung des Rauches der Silberhütten auf die benachbarte Vegetation. *Berg- und hüttenmännische Zeitung* 9, 305–308, 327–31, 344–8, 361–6.

Stoixner, L.v., 1788. *Zufällige Gedanken von dem Holzmangel.* München.

Stowe, J. 1632. *Annales or Generall Chronicle of England.* London

Tainter, J.A., 1988. *The Collapse of Complex Societies.* Cambridge.

Tann, J., 1973. Fuel Saving in the Process Industries during the Industrial Revolution. *Business History* 15, 149–159.

Te Brake, W.H., 1975. Air Pollution and Fuel Crises in Preindustrial London, 1250–1650. *Technology and Culture* 16, 337–359.

Temin, P., 1966. Steam and Waterpower in the Early 19th Century. *Journal of Economic History* 26, 187–205.

Thirring, H., 1958. *Energy for Man. From Windmills to Nuclear Power.* Bloomington.

Thomas, B., 1980. Towards an Energy Interpretation of the Industrial Revolution. *Atlantic Economic Journal* 8, 1–13.

Thon, F., 1832. Art. Holzsparkunst. In: Ersch, J.S. and Gruber, J.G. (eds) *Allgemeine Encyclopaedie der Wissenschaften und Künste*, sect. 2, vol. 9, 190–3. Leipzig.

Timm, A., 1960. *Die Waldnutzung in Nordwestdeutschland im Spiegel der Weistümer.* Köln and Graz.

Tooby, John and Cosmides, Leda, 1992. The Psychological Foundations of Culture. In: Barkow, Jerome H.: Cosmides, Leda and Tooby, John (eds), *The Adapted Mind. Evolutionary Psychology and the Generation of Culture*, 19–136. New York.

Tunzelmann, G.N.v., 1978. *Steam Power and British Industrialization to 1860.* Oxford.

Varchmin, J. and Radkau, J., 1979. *Kraft, Energie, Arbeit.* München.

Venel, D., 1780. *Unterricht von den Steinkohlen und ihrem Gebrauche zu allen Arten von Feuern.* Dresden.

Voigt, J.C.W., 1802/05. *Versuch einer Geschichte der Steinkohlen, der Braunkohlen und des Torfs*, 2 vols. Weimar.

Von der Steinkohlenfeuerung in Schlesien. *Schlesische Provinzialblätter* 1786, Vol. 2, 221ff.

Wackernagel, M. and Rees, W., 1996. *Our Ecological Footprint.* Gabriola Island, B.C.

Wagner, E., 1930. *Die Holzversorgung der Lüneburger Saline.* Düsseldorf.

Wallerstein, Immanuel 1974. *The Modern World System*, Vol. 1: *Capitalist Agriculture and the Origins of the European World-Economy in the 16th Century.* New York.

Webb, Walter Prescott 1952. *The Great Frontier.* Boston.

Weber, M., 1909. „Energetische" Kulturtheorien. *Archiv für Sozialwissenschaft und Sozialpolitik* 29, 575–98.

Weber, W., 1978. Die Schiffbarmachung der Ruhr und die Industrialisierung im Ruhrgebiet. In: Kellenbenz, H. (ed.), 1978. *Wirtschaftswachstum, Energie und Verkehr vom Mittelalter bis ins 19. Jahrhundert*, 95–115. Stuttgart and New York.

West, E., 1815. *Essay on the Application of Capital to Land.* London.

White, Leslie A. 1943. Energy and the Evolution of Culture. *American Anthropologist* 45, 335–56.

White, Leslie A. 1949. *The Science of Culture.* New York.

White, Leslie A. 1954. The Energy Theory of Cultural Development. In: Kapadia, K.M. (ed.), *Professor Ghurye Felicitation Volume*, 1–10. Bombay.

White, Leslie A. 1959. *The Evolution of Culture. The Development of Civilization to the Fall of Rome.* New York.

Wilkinson, R.G., 1973. *Poverty and Progress.* London.

Williams, D.M., 1966. Merchanting in the First Half of the 19th Century. The Liverpool Timber Trade. *Business History* 8, 103–117.

Williams, J., 1789. *Natural History of the Mineral Kingdom*, Vol. 1. Edinburgh.

Wilsdorf, H., 1960. Holz, Erz, Salz. Das Transportproblem im Montanwesen. In: *Bergbau, Wald, Flöße. Freiberger Forschungshefte* D 28. Berlin.

Winiwarter, V., 1999. Böden in Agrargesellschaften: Wahrnehmung, Behandlung und Theorie von Cato bis Palladius. In: Sieferle, R.P. and Breuninger, H. (ed.), *Natur-Bilder. Wahrnehmungen von Natur und Umwelt in der Geschichte*, 181–221. Frankfurt am Main.

Wrigley, E.A., 1962. The Supply of Raw Materials in the Industrial Revolution. *Economic History Review* 15, 1–16.

Wrigley, E.A., 1988. *Continuity, Chance and Change. The Character of the Industrial Revolution in England*. Cambridge.

Wrigley, E.A., 1994. The Classical Economists, the Stationary State, and the Industrial Revolution. In: G.D. Snooks (ed.), *Was the Industrial Revolution Necessary?*, 27–42. London.

Yarranton, A., 1677. *England's Improvement by Sea and Land to Out-do the Dutch without Fighting, to Pay Debts without Moneys, to set at Work all the Poor of England with the Growth of our own Lands*. London.

Yoffee, N. and Cowgill, G.L. (eds), 1988. *The Collapse of Ancient States and Civilizations*. Tucson.

Zedler, J.H., 1732–1750. *Großes vollständiges Universallexicon aller Wissenschaften und Künste*, 64 vols. Halle and Leipzig.

Zeeuw, J.W. de 1978. Peat and the Dutch Golden Age. The Historical Meaning of Energy Attainability. *Afdeling Agrarische Geschiedenis*, Bijdragen 21, 3–31.

Zeiller, M., 1634. *Itinerarium Magnae Britanniae oder Raißbeschreibung durch Engell-Schott- und Irrland*. Straßburg.

Index